Everyday Mathematics®

Student Math Journal 1

The University of Chicago
School Mathematics Project

The McGraw-Hill Companies

UCSMP Elementary Materials Component

Max Bell, Director

Authors

Max Bell
John Bretzlauf
Amy Dillard
Robert Hartfield
Andy Isaacs
James McBride, Director
Kathleen Pitvorec
Peter Saecker
Robert Balfanz*
William Carroll*
Sheila Sconiers*

Technical Art

Diana Barrie

First Edition only

Contributors

Martha Ayala, Virginia J. Bates, Randee Blair, Donna R. Clay, Vanessa Day, Jean Faszholz, James Flanders, Patti Haney, Margaret Phillips Holm, Nancy Kay Hubert, Sybil Johnson, Judith Kiehm, Carla L. LaRochelle, Deborah Arron Leslie, Laura Ann Luczak, Mary O'Boyle, William D. Pattison, Beverly Pilchman, Denise Porter, Judith Ann Robb, Mary Seymour, Laura A. Sunseri

This material is based upon work supported by the National Science Foundation under Grant No. ESI-9252984. Any opinions, findings, and conclusions or recommendations expressed in this material are those of the authors and do not necessarily reflect the views of the National Science Foundation.

www.sra4kids.com

SRA

Send all inquiries to:
SRA/McGraw-Hill
P.O. Box 812960
Chicago, IL 60681

Printed in the United States of America.

ISBN 0-07-600011-7

14 15 16 DBH 10 09 08 07

The McGraw-Hill Companies

Contents

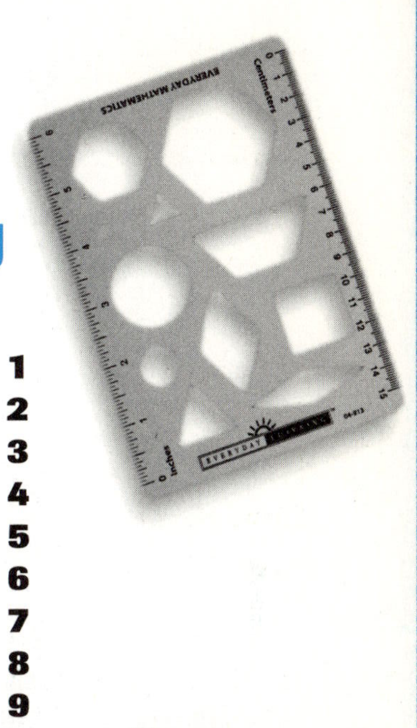

Unit 1: Naming and Constructing Geometric Figures

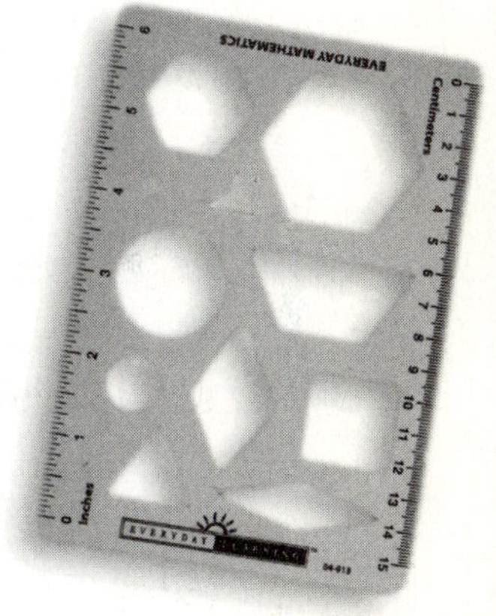

A note at the bottom of each journal page indicates when that page is first used.
Some pages will be used again during the course of the year.

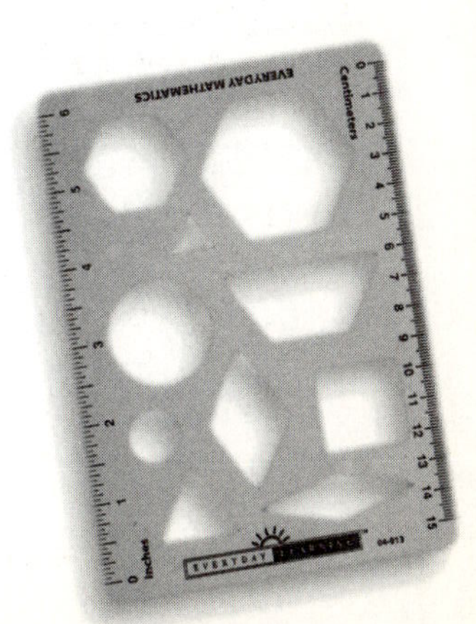

Unit 2: Using Numbers and Organizing Data

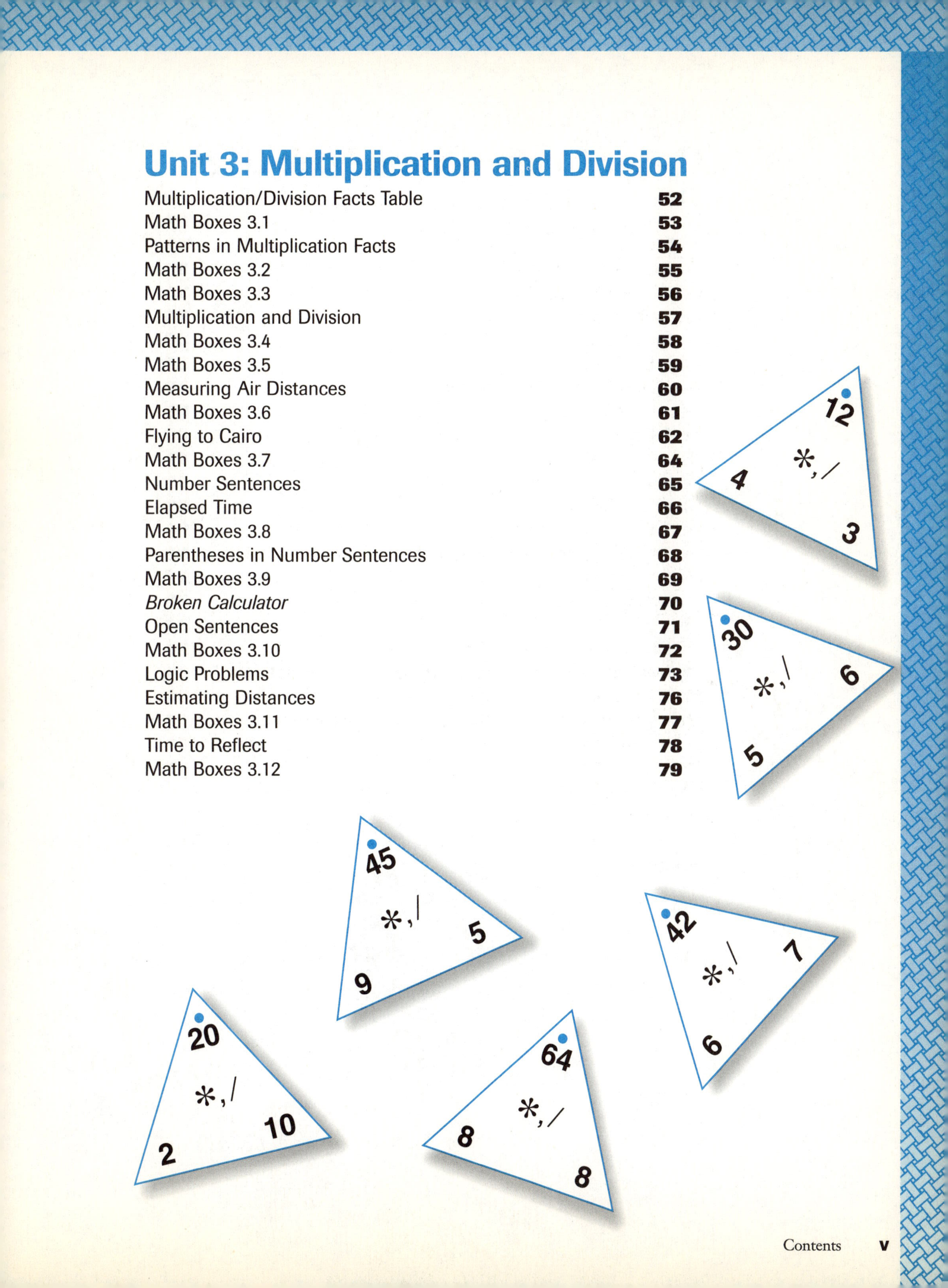

Unit 3: Multiplication and Division

Unit 4: Decimals and Their Uses

Unit 5: Big Numbers, Estimation, and Computation

Unit 6: Division; Map Reference Frames; Measures of Angles

Activity Sheets

Date Time

Welcome to *Fourth Grade Everyday Mathematics*®

Much of your work in kindergarten through third grade was basic training in mathematics and its uses. You spent quite a bit of time and effort telling and solving number stories and learning arithmetic, including the basic addition and multiplication facts.

Fourth Grade Everyday Mathematics builds on this basic training and begins to make the transition to mathematics concepts and ways of using mathematics that are more like what your parents and older siblings may have learned in high school. We believe, along with many other people, that fourth graders today can learn more and do more than was thought to be possible ten or twenty years ago.

Here are some things you will be asked to do in *Fourth Grade Everyday Mathematics:*

- Strengthen and increase your "number sense," "measure sense," and estimation skills.
- Review and extend your skills in the basics of arithmetic—addition, subtraction, multiplication, and division. There isn't much more to learn about the arithmetic of whole numbers, but over the next couple of years you will become comfortable with the many uses of fractions, percents, and decimals.
- Learn about "variables" (letters that stand for numbers) and other topics in algebra that your parents or brothers and sisters may not have learned until they were in high school.
- Develop your geometry skills and concepts further, with more exact definitions and classifications of geometric figures, with constructions and transformations of figures, and with more work on areas of 2-dimensional figures and volumes of 3-dimensional shapes.
- Take a World Tour. Along the way you will consider many kinds of data about various countries and learn about coordinate systems used to locate places on world globes and on flat maps.
- Do many projects involving numerical data.

In fourth grade, you will be asked to do more independent reading and to rely on what you can find out for yourself (often working with partners or in groups) rather than being told everything by your teacher.

The authors hope that you find the activities fun and that you see some of the real beauty in mathematics. And most importantly, they hope you become better and better at using mathematics to sort out and solve interesting problems in your life.

Date Time

Using Your *Student Reference Book*

Use your *Student Reference Book* to complete the following:

1. Look up the word **mode** in the Glossary.

a. Copy the definition. ______________________

____

b. On which page in the *Student Reference Book* could you find more information about the mode of a set of data? ______________________

2. Read the essay "Comparing Numbers and Amounts."

a. Solve the Check Your Understanding problems.

Problem 1: __________ Problem 2: __________

Problem 3: __________ Problem 4: __________

b. Check your answers in the Answer Key.

c. Describe what you did to find the essay.

____

____

3. Look up the rules for the game *Beat the Calculator.*

a. On which page did you find the rules? ______________________

b. How many players are needed for the game? ______________________

4. Find the World Tour section. Record two interesting facts you find there.

a. Fact 1: ______________________

I found this information on page ______________________.

b. Fact 2: ______________________

I found this information on page ______________________.

Date Time

Math Boxes 1.1

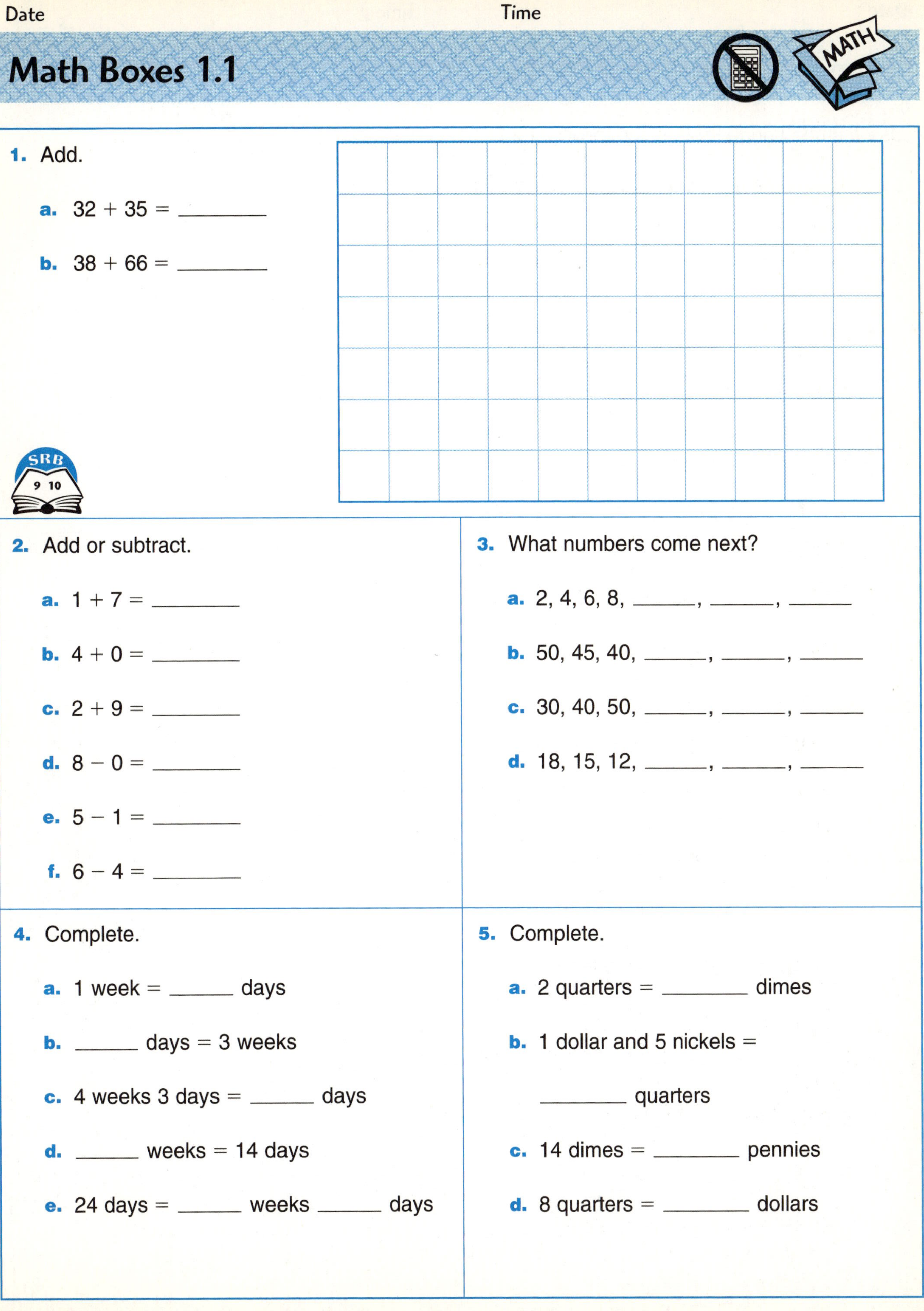

1. Add.

a. 32 + 35 = ______

b. 38 + 66 = ______

2. Add or subtract.

a. 1 + 7 = ______

b. 4 + 0 = ______

c. 2 + 9 = ______

d. 8 − 0 = ______

e. 5 − 1 = ______

f. 6 − 4 = ______

3. What numbers come next?

a. 2, 4, 6, 8, ______, ______, ______

b. 50, 45, 40, ______, ______, ______

c. 30, 40, 50, ______, ______, ______

d. 18, 15, 12, ______, ______, ______

4. Complete.

a. 1 week = ______ days

b. ______ days = 3 weeks

c. 4 weeks 3 days = ______ days

d. ______ weeks = 14 days

e. 24 days = ______ weeks ______ days

5. Complete.

a. 2 quarters = ______ dimes

b. 1 dollar and 5 nickels =

______ quarters

c. 14 dimes = ______ pennies

d. 8 quarters = ______ dollars

Date Time

Points, Line Segments, Lines, and Rays

Use a straightedge to draw the following:

1. a. Draw and label line segment *TO* ($\overline{TO}$).

   b. What is another name for $\overline{TO}$? ________

2. a. Draw and label line *IF* ($\overleftrightarrow{IF}$). Draw and label point *T* on it.

   b. What are two other names for $\overleftrightarrow{IF}$? ____________________

3. a. Draw and label ray *ON* ($\overrightarrow{ON}$). Draw and label point *R* on it.

   b. What is another name for $\overrightarrow{ON}$? ________

4. a. Below, draw a line segment from each point to each of the other points.

M *N*

O *P*

   b. How many line segments did you draw? ________

   c. Write a name for each line segment you drew.

____

____

Date

Math Boxes 1.2

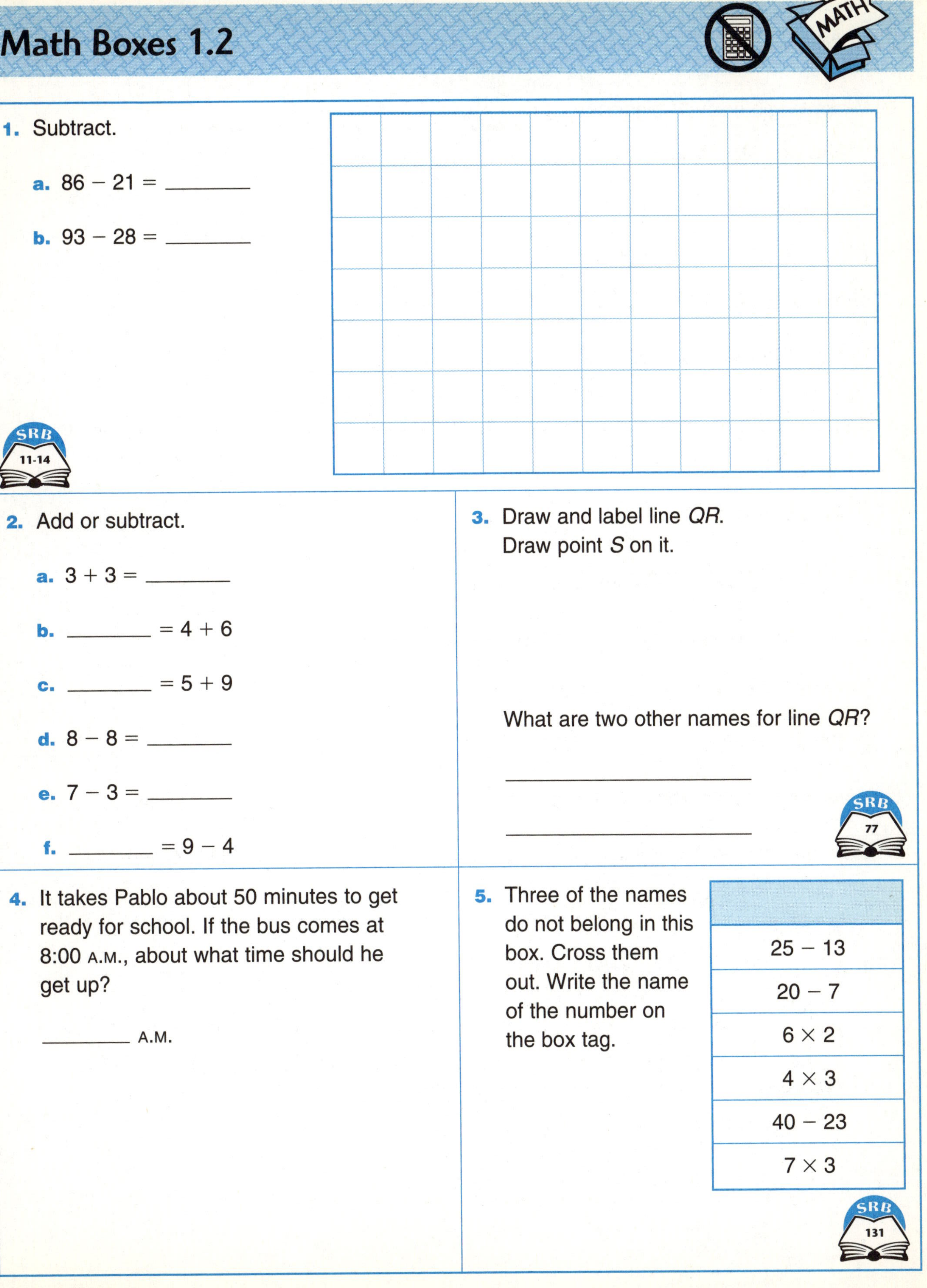

1. Subtract.

   a. 86 − 21 = ______

   b. 93 − 28 = ______

2. Add or subtract.

   a. 3 + 3 = ______

   b. ______ = 4 + 6

   c. ______ = 5 + 9

   d. 8 − 8 = ______

   e. 7 − 3 = ______

   f. ______ = 9 − 4

3. Draw and label line *QR*.
   Draw point *S* on it.

   What are two other names for line *QR*?

   ______

   ______

4. It takes Pablo about 50 minutes to get ready for school. If the bus comes at 8:00 A.M., about what time should he get up?

   ______ A.M.

5. Three of the names do not belong in this box. Cross them out. Write the name of the number on the box tag.

25 − 13
20 − 7
6 × 2
4 × 3
40 − 23
7 × 3

Date Time

Angles

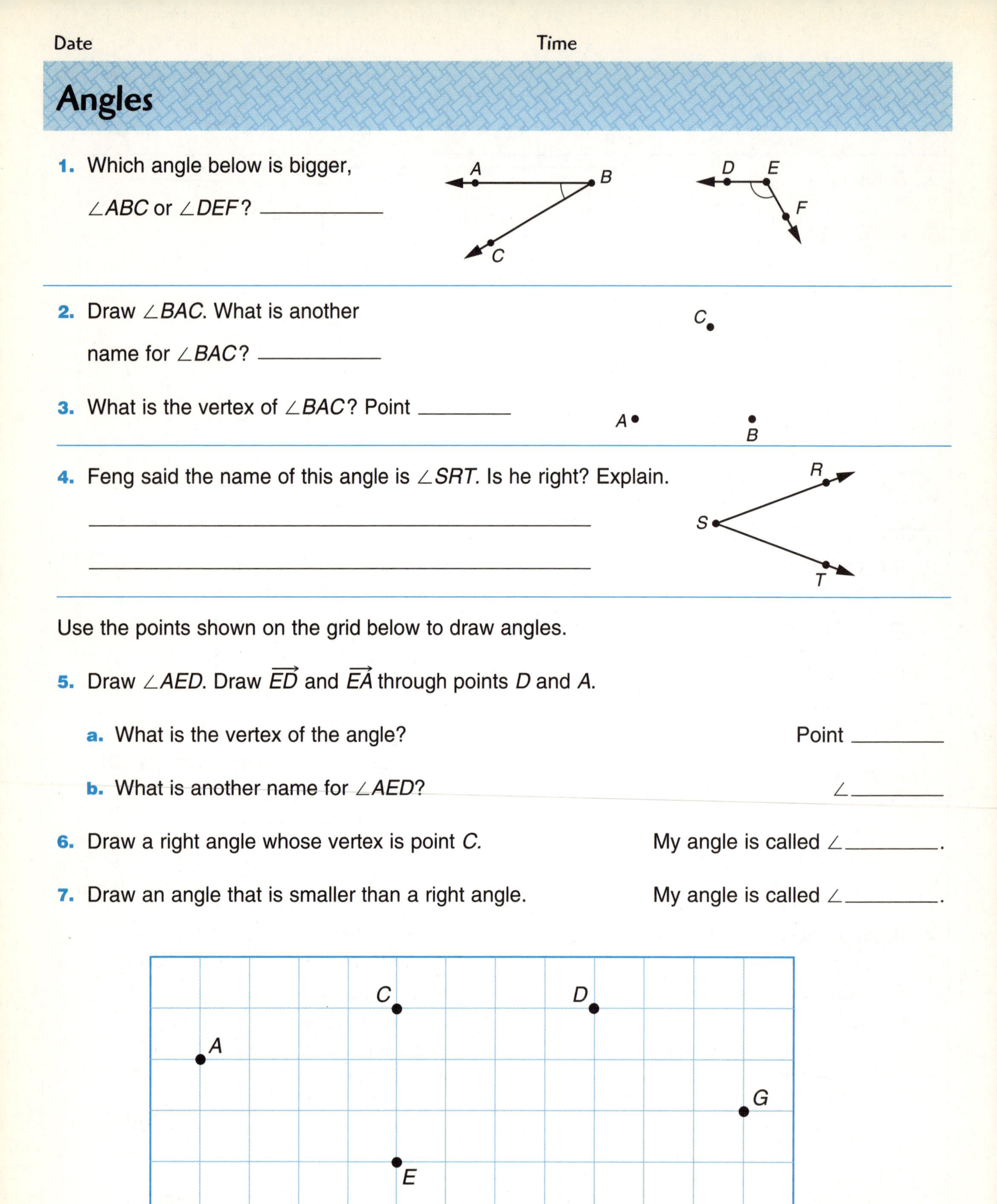

1. Which angle below is bigger, $\angle ABC$ or $\angle DEF$? ________

2. Draw $\angle BAC$. What is another name for $\angle BAC$? ________

3. What is the vertex of $\angle BAC$? Point ________

4. Feng said the name of this angle is $\angle SRT$. Is he right? Explain.

____

____

Use the points shown on the grid below to draw angles.

5. Draw $\angle AED$. Draw $\overrightarrow{ED}$ and $\overrightarrow{EA}$ through points D and A.

a. What is the vertex of the angle? Point ________

b. What is another name for $\angle AED$? $\angle$ ________

6. Draw a right angle whose vertex is point C. My angle is called $\angle$ ________.

7. Draw an angle that is smaller than a right angle. My angle is called $\angle$ ________.

Date Time

Math Boxes 1.3

MATH

1. Add.

   a. 63 + 12 = ______

   b. 56 + 97 = ______

2. Add or subtract.

   a. 2 + 5 = ______

   b. 6 + 8 = ______

   c. 9 + 3 = ______

   d. ______ = 12 − 5

   e. ______ = 15 − 9

   f. ______ = 11 − 7

3. What numbers come next?

   a. 4, 8, 12, 16, ______, ______, ______

   b. 33, 30, 27, ______, ______, ______

   c. 20, 35, 50, ______, ______, ______

   d. 37, 31, 25, ______, ______, ______

4. Complete.

   a. 2 weeks = ______ days

   b. 11 weeks = ______ days

   c. 5 weeks 4 days = ______ days

   d. 21 days = ______ weeks

   e. 30 days = ______ weeks ______ days

5. Complete.

   a. 4 quarters = ______ dimes

   b. 2 dollars and 10 nickels = ______ quarters

   c. 9 dimes = ______ pennies

   d. 30 dimes = ______ dollars

Date Time

Parallelograms

1. Circle the pairs of line segments below that are parallel. Check your answers by extending each pair of segments to see if the two segments in the pair meet.

a. **b.** **c.** **d.**

Use your Geometry Template or straightedge to draw the following quadrangles:

2. Draw a quadrangle that has two pairs of parallel sides.

This is called a ________________.

3. Draw a quadrangle that has only one pair of parallel sides.

This is called a ________________.

4. A **parallelogram** is a quadrangle that has two pairs of parallel sides. Which of the following quadrangles are parallelograms?

a. squares **b.** rectangles **c.** rhombuses **d.** trapezoids

5. A **rhombus** is a parallelogram in which all sides are the same length. Which of the following quadrangles are always rhombuses?

a. squares **b.** rectangles **c.** trapezoids

6. A **rectangle** is a parallelogram that has all right angles. Which of the following are rectangles? Write **always, sometimes,** or **never** to complete each sentence.

a. **Squares** are ________________ rectangles.

b. **Rhombuses** are ________________ rectangles.

c. **Trapezoids** are ________________ rectangles.

7. Is a kite a parallelogram? ____________ Explain below.

____

____

Date Time

Math Boxes 1.4

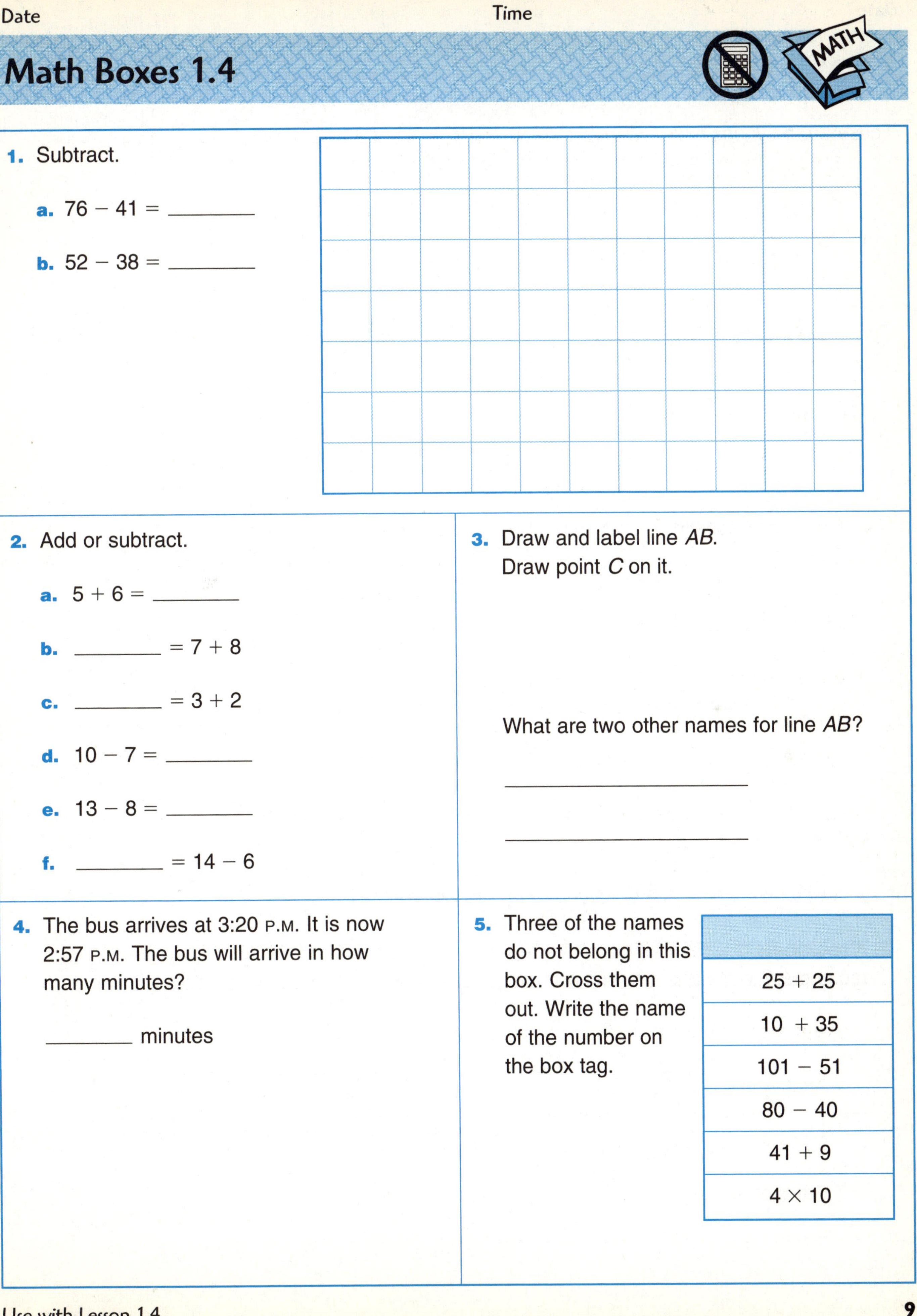

1. Subtract.

   a. $76 - 41 =$ ______

   b. $52 - 38 =$ ______

2. Add or subtract.

   a. $5 + 6 =$ ______

   b. ______ $= 7 + 8$

   c. ______ $= 3 + 2$

   d. $10 - 7 =$ ______

   e. $13 - 8 =$ ______

   f. ______ $= 14 - 6$

3. Draw and label line *AB*. Draw point *C* on it.

   What are two other names for line *AB*?

   ______

   ______

4. The bus arrives at 3:20 P.M. It is now 2:57 P.M. The bus will arrive in how many minutes?

   ______ minutes

5. Three of the names do not belong in this box. Cross them out. Write the name of the number on the box tag.

$25 + 25$
$10 + 35$
$101 - 51$
$80 - 40$
$41 + 9$
4×10

Date Time

What Is a Polygon?

These are polygons.

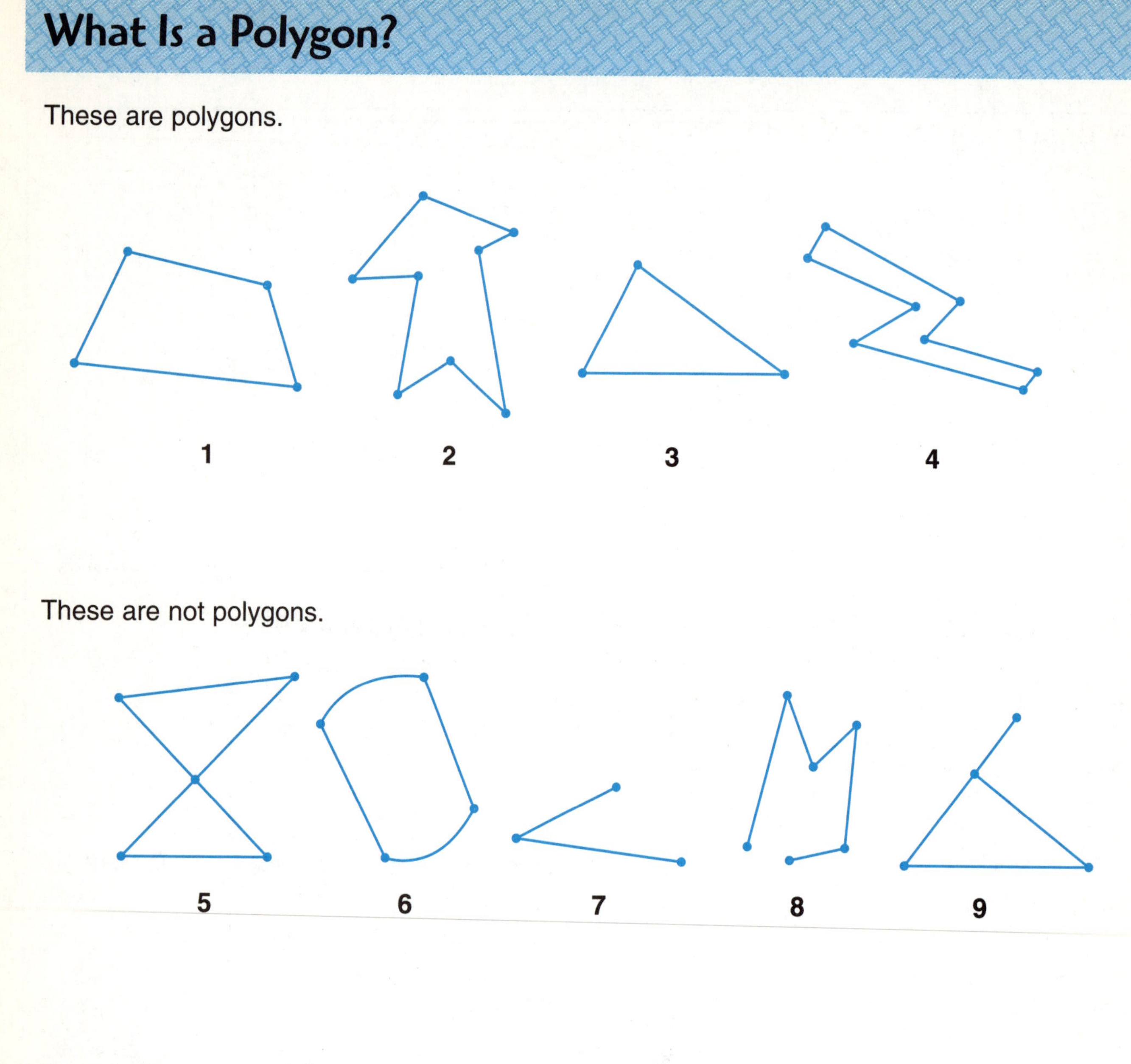

These are not polygons.

If you had to explain what a polygon is, what would you say?

____

____

____

____

____

____

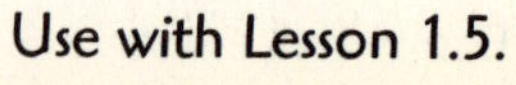

Date Time

Math Boxes 1.5

MATH

1. Put these numbers in order from smallest to largest.

10,005 5,100

51,000 10,500

____

____

____

____

2. Draw $\angle ART$. What is another name for $\angle ART$? ________

T

A

R

SRB 78

3. Add or subtract.

a. $6 + 3 =$ ________

b. $1 + 4 =$ ________

c. ________ $= 9 + 9$

d. ________ $= 10 - 8$

e. $13 - 7 =$ ________

f. $16 - 9 =$ ________

4. Draw and label line segment AB.

SRB 76

5. Name as many rays as you can in the figure below.

M N O

Write their names.

____

____

SRB 77

6. Draw a quadrangle with 2 pairs of parallel sides.

What kind of quadrangle is this?

____

SRB 85

Date Time

An Inscribed Square

Follow the directions below to make a square.

Step 1 Use your compass to draw a circle on a sheet of paper. The circle should be small enough to fit on the next page. Cut out the circle.

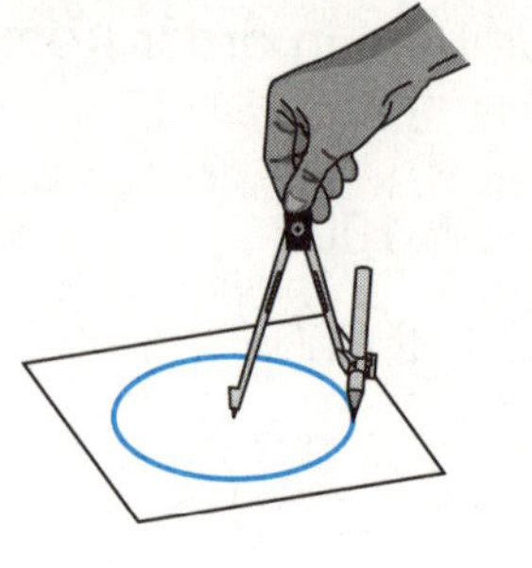

Step 2 With your pencil, make a dot in the center of the circle, where the hole is, on both the front and the back.

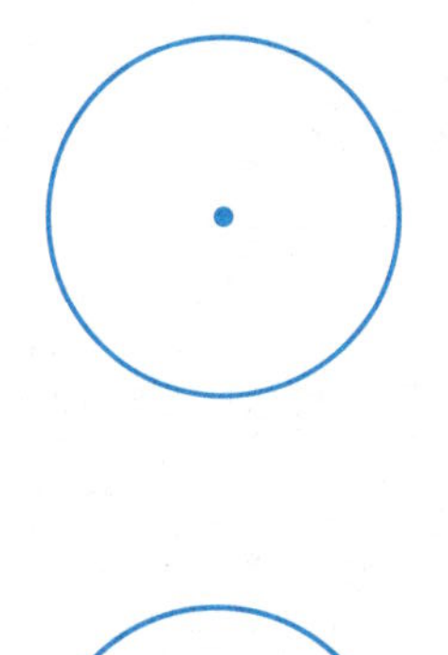

Step 3 Fold the circle in half. Make sure that the edges match and that the fold line passes through the center.

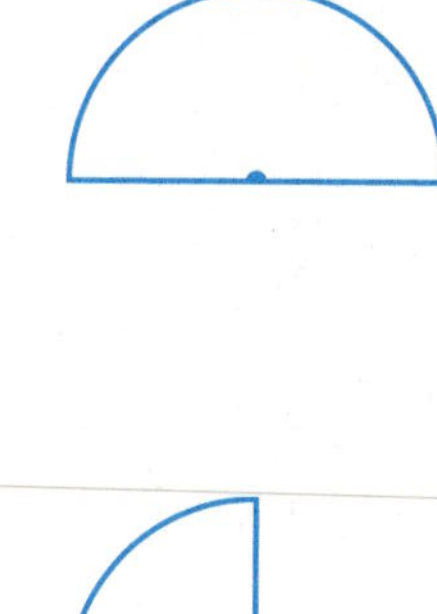

Step 4 Fold the folded circle in half again so that the edges match.

Step 5 Unfold your circle. The folds should pass through the center of the circle and form four right angles.

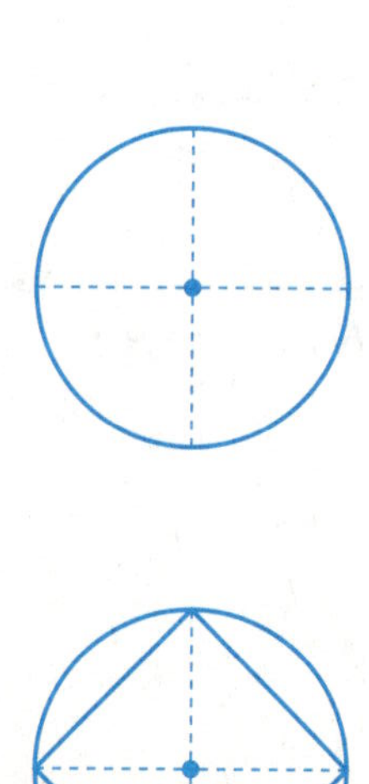

Step 6 Connect the endpoints of the folds with a straightedge to make a square. Cut out the square.

Use with Lesson 1.6.

Date Time

An Inscribed Square (cont.)

Now use your compass to find out whether the four sides of your square are all about the same length. If they are, paste or tape the square in the space below. If not, follow the directions on page 12 again. Paste or tape your second square in the space below.

Date Time

Math Boxes 1.6

1. In the numeral 42,018, the 2 stands for 2,000.

   a. The 1 stands for ________.

   b. The 8 stands for ________.

   c. The 4 stands for ________.

SRB 4

2. Which of the shape(s) below are **not** polygons? ____

A B C

SRB 82

3. Add or subtract.

   a. $6 + 7 =$ ____

   b. $8 + 8 =$ ____

   c. ____ $= 9 + 4$

   d. ____ $= 12 - 9$

   e. $14 - 7 =$ ____

   f. $13 - 5 =$ ____

4. Draw a quadrangle that does not have any right angles.

SRB 86

5. Circle the convex polygons.

SRB 83

6. Draw and label ray *HA*. Draw point *T* on it.

   What is another name for ray *HA*?

   ____

SRB 77

Date Time

Circle Constructions

Do each of the following 3 constructions on a separate sheet of paper. Try and try again until you are satisfied with your work. Then cut out your 3 best constructions and paste them in your journal.

1. Use your compass to draw a picture of a circular dartboard. Paste your best work in the space below.

The circles in your picture are called **concentric circles.**

Date Time

Circle Constructions (cont.)

2. a. Make a dot near the center of your paper. Use your compass to draw a circle with that dot as its center.

b. *Without changing the opening of your compass,* draw a second circle that passes through the center of the first circle. Mark the center of the second circle.

c. *Without changing the opening of your compass,* draw a third circle that passes through the center of each of the first two circles.

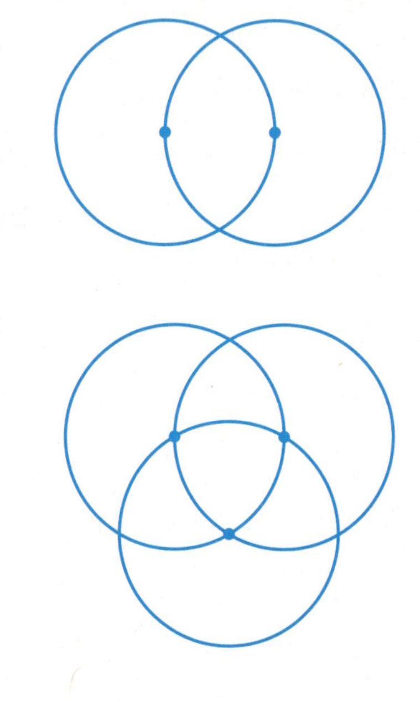

Try and try again until you are satisfied with your work. Then cut out your circle design and paste it in the space below.

Date Time

Circle Constructions (cont.)

Challenge

3. Try to draw this design with your compass. Work on separate sheets of paper until you are satisfied with your work. Color your best design. Then cut it out and paste it in the space below.

   Hint: Start by making the 3-circle design on page 16. Then add more circles to it.

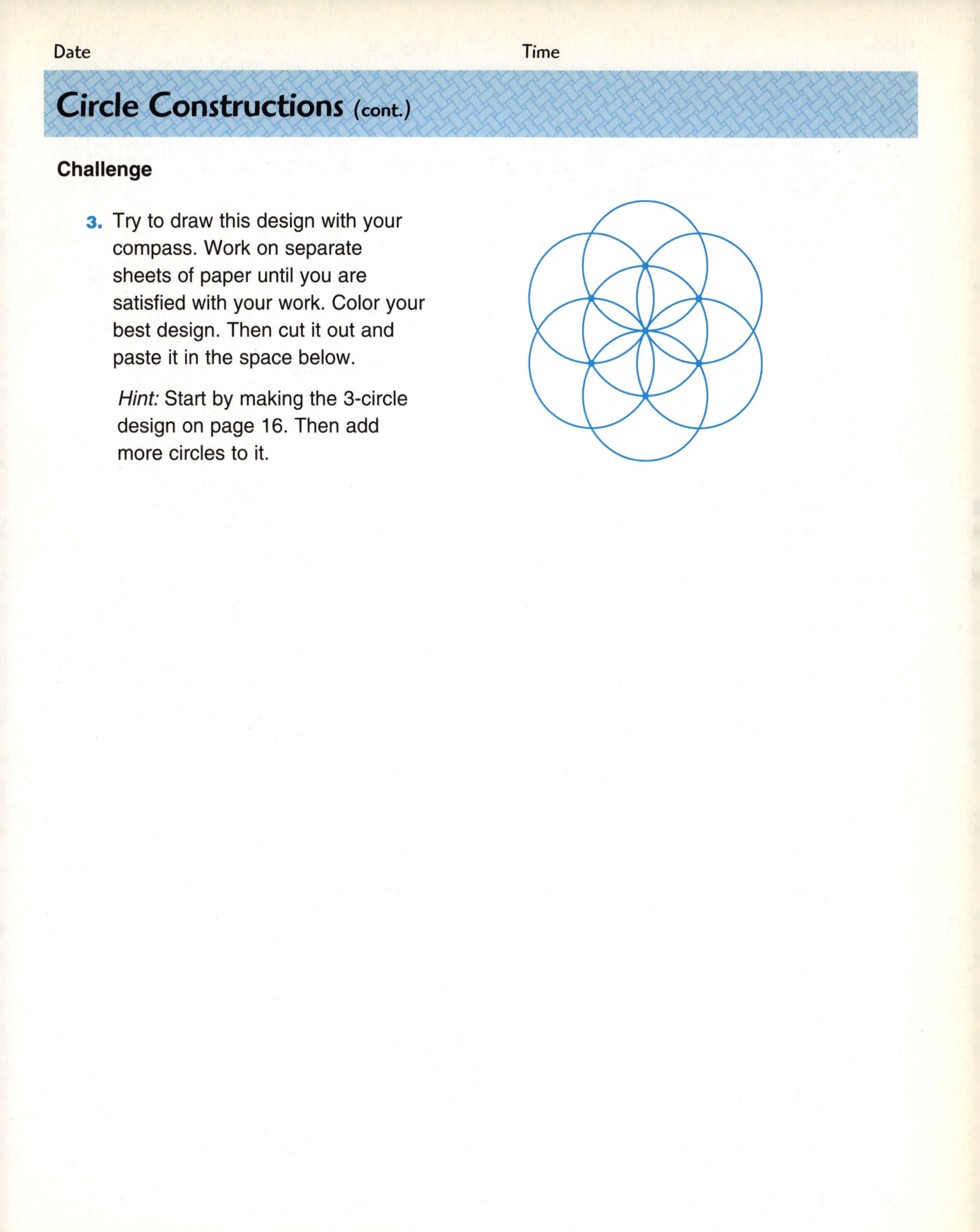

Date ____ Time ____

Polygon Checklist

Place a check mark next to all of the statements that are true about each figure. Write an additional true statement for that figure.

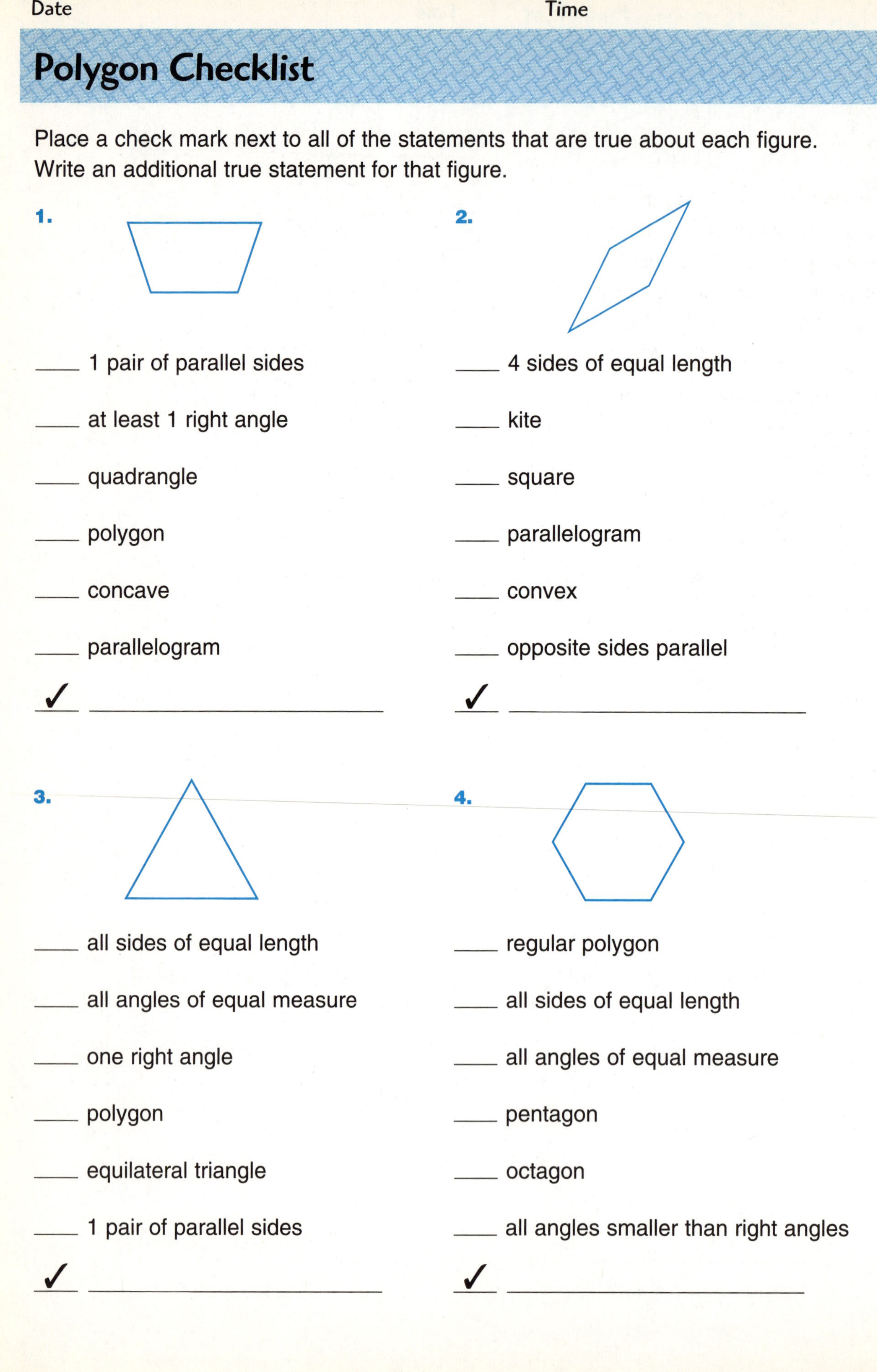

1.

____ 1 pair of parallel sides

____ at least 1 right angle

____ quadrangle

____ polygon

____ concave

____ parallelogram

✓ ____________________

2.

____ 4 sides of equal length

____ kite

____ square

____ parallelogram

____ convex

____ opposite sides parallel

✓ ____________________

3.

____ all sides of equal length

____ all angles of equal measure

____ one right angle

____ polygon

____ equilateral triangle

____ 1 pair of parallel sides

✓ ____________________

4.

____ regular polygon

____ all sides of equal length

____ all angles of equal measure

____ pentagon

____ octagon

____ all angles smaller than right angles

✓ ____________________

Date Time

Math Boxes 1.7

MATH

1. Put these numbers in order from smallest to largest.

32,000 3,200

23,000 2,300

______

______

______

______

2. Draw $\angle TIF$. What is the vertex of $\angle TIF$? ______

F T I

3. Add or subtract.

a. $8 + 3 =$ ______

b. $6 + 6 =$ ______

c. ______ $= 5 + 7$

d. ______ $= 13 - 9$

e. ______ $= 11 - 4$

f. $10 - 5 =$ ______

4. Draw and label line segment *GP*.

5. Name as many rays as you can in the figure below.

L M N O P

Write their names.

______

______

6. Draw a quadrangle with 1 pair of parallel sides.

What kind of quadrangle is this?

______

Copying a Line Segment

Steps 1–4 below show you how to copy a line segment.

Step 1 You are given line segment *AB* to copy.

Step 2 Draw a line segment longer than line segment *AB*. Label one of its endpoints *C*.

Step 3 Open your compass so that the anchor is on one endpoint of line segment *AB* and the pencil point is on the other endpoint.

Step 4 *Without changing the compass opening,* place the anchor on point *C* on your second line segment. Make a mark that crosses this line segment. Label the point where the mark crosses the line segment with the letter *D*.

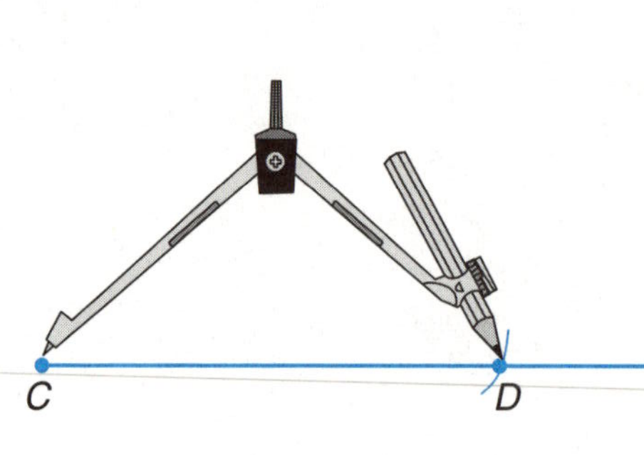

Line segment *CD* should be about the same length as line segment *AB*.

Use a compass and straightedge to copy the line segments shown below.

Constructing a Regular, Inscribed Hexagon

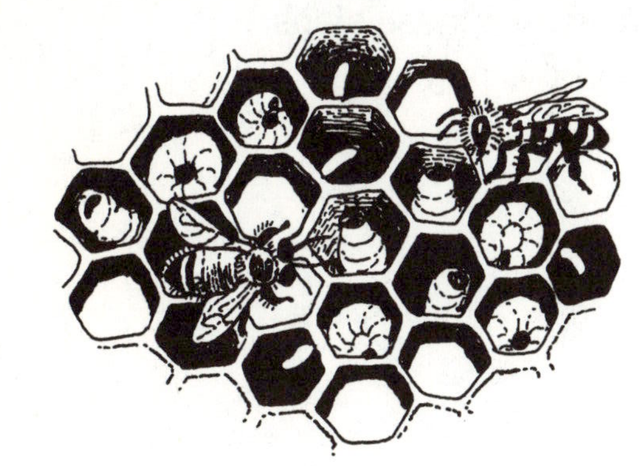

Hexagons are seen in the natural world and in things that people make and use. For example, honeycombs are built with many hexagonal shapes, and snowflakes suggest the shape of a hexagon.

Soccer balls are made up of regular hexagons and regular pentagons.

For many centuries, wonderful tile designs have been created all over the world. Some of the best tile designs are from Islamic art. The designs often start with hexagons, which fit together without gaps or overlaps. As these pictures show, tile designs can be developed in many ways from a pattern that uses hexagons.

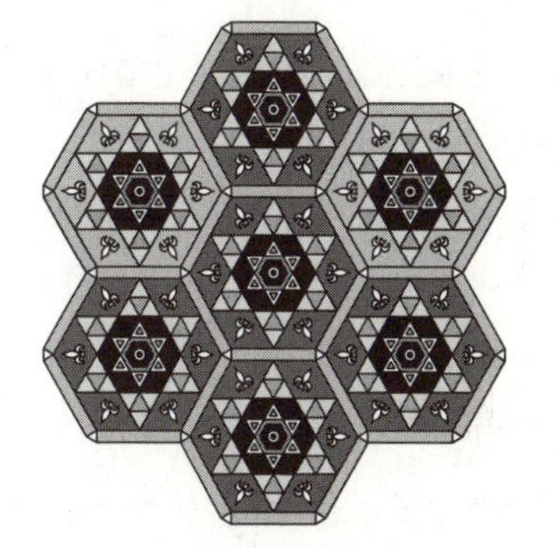

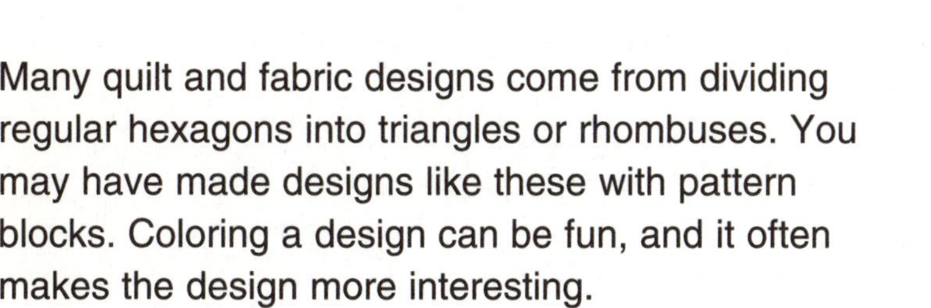

Many quilt and fabric designs come from dividing regular hexagons into triangles or rhombuses. You may have made designs like these with pattern blocks. Coloring a design can be fun, and it often makes the design more interesting.

Date Time

Constructing a Regular, Inscribed Hexagon (cont.)

Follow each step below. Draw on a separate sheet of paper. Repeat these steps several times. Tape or paste your best work onto the bottom of this page.

Step 1 Draw a circle. (Keep the same compass opening for Steps 2 and 3.) Draw a dot on the circle. Place the anchor of your compass on the dot and draw a mark on the circle.

Step 2 Place the anchor of your compass on the mark you just made and draw another mark on the circle.

Step 3 Do this four more times to divide the circle into 6 equal parts. The 6th mark should be on the dot you started with or very close to it.

Step 4 With your straightedge, connect the 6 marks on the circle to form a regular hexagon. Use your compass to check that the sides of the hexagon are all about the same length.

Date Time

More Constructions

Construct a regular hexagon on a separate sheet of paper. Then divide the hexagon into 6 equilateral triangles.

Try this several times until you are satisfied with your work. Tape or paste your best work in the space below.

Date Time

Math Boxes 1.8

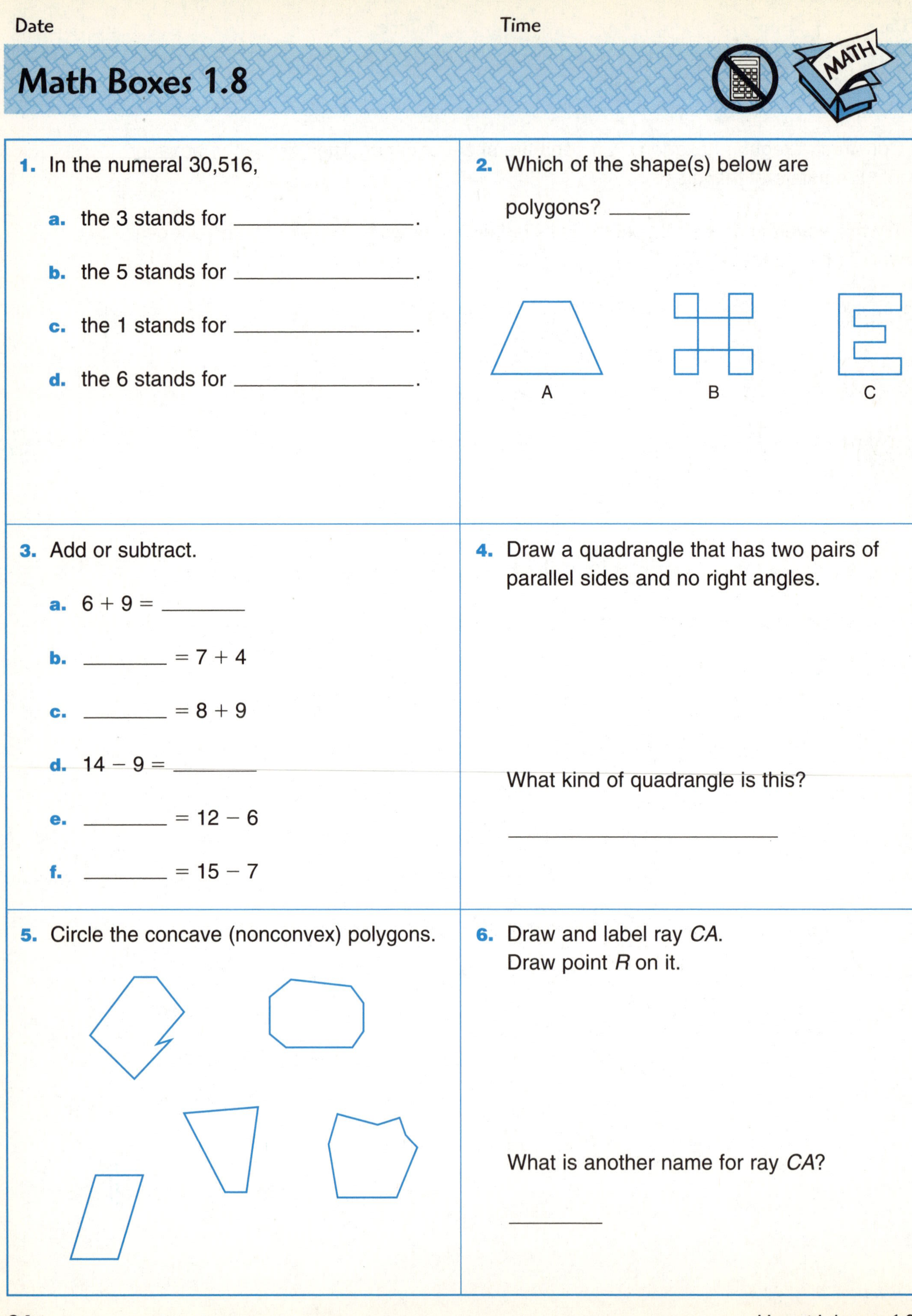

1. In the numeral 30,516,

   a. the 3 stands for ________________.

   b. the 5 stands for ________________.

   c. the 1 stands for ________________.

   d. the 6 stands for ________________.

2. Which of the shape(s) below are polygons? ______

3. Add or subtract.

   a. $6 + 9 =$ ______

   b. ______ $= 7 + 4$

   c. ______ $= 8 + 9$

   d. $14 - 9 =$ ______

   e. ______ $= 12 - 6$

   f. ______ $= 15 - 7$

4. Draw a quadrangle that has two pairs of parallel sides and no right angles.

   What kind of quadrangle is this?

   ____

5. Circle the concave (nonconvex) polygons.

6. Draw and label ray *CA*. Draw point *R* on it.

   What is another name for ray *CA*?

   ____

Date Time

Time to Reflect

1. Write 2 or 3 interesting things you have learned about geometry.

2. What do you think the difference is between *drawing* a geometric figure and *constructing* a geometric figure?

3. What was the hardest part of this unit for you?

Date Time

Math Boxes 1.9

1. Add.

a. $\begin{array}{r} 64 \\ +\ 32 \\ \hline \end{array}$

b. $\begin{array}{r} 48 \\ +\ 96 \\ \hline \end{array}$

2. Subtract.

a. $\begin{array}{r} 78 \\ -\ 42 \\ \hline \end{array}$

b. $\begin{array}{r} 81 \\ -\ 36 \\ \hline \end{array}$

3. Put these numbers in order from smallest to largest.

46,000 40,600

4,600 4,006

____

____

____

____

4. In the numeral 78,965,

a. the 8 stands for __________.

b. the 6 stands for __________.

c. the 7 stands for __________.

d. the 9 stands for __________.

5. Use the following list of numbers to answer the questions:

12, 3, 15, 6, 12, 14, 6, 5, 9, 12

a. Which number is the smallest? ______

b. Which number is the largest? ______

c. What is the difference between the smallest and largest numbers? ______

d. Which number appears most often? ______

Date Time

A Visit to Washington, D.C.

Refer to pages 211 and 212 in your *Student Reference Book.*

1. About how many people tour the White House every year?
Check the best answer.

____ between 100 thousand and 1 million ____ between 1 million and 10 million

____ between 10 million and 100 million ____ between 100 million and 1 billion

2. About how many people ride the Washington Metrorail on an average weekday?
Check the best answer.

____ between 100 thousand and 1 million ____ between 1 million and 10 million

____ between 10 million and 100 million ____ between 100 million and 1 billion

3. The Library of Congress adds about __________ items each day. About how many days does it take to add 50,000 items to the Library of Congress? __________________

4. Write the year that each event happened. Then draw a dot for each event on the timeline below. Label the dot with the correct letter and date.

A The year the Metrorail opened 1976

B The year of the flight of the *Flyer* __________

C The year the Washington Monument was completed __________

D The year the Lincoln Memorial was dedicated __________

E The year of the first nonstop flight across the Atlantic __________

F The year the Jefferson Memorial was dedicated __________

G The year of the first landing on the moon __________

1850 1900 1950 A 1976 2000

Date Time

Math Boxes 2.1

1. Complete the division facts.

a. $28 \div 4 =$ ______

b. $27 \div$ ______ $= 9$

c. $42 \div$ ______ $= 6$

d. $24 \div$ ______ $= 6$

e. $36 \div 6 =$ ______

f. $90 \div 9 =$ ______

SRB 132 152

2. A giant tortoise can live about 150 years. An elephant can live about 78 years. About how much longer can a giant tortoise live than an elephant?

About ______ years

3. What is the value of the digit 7 in the numerals below?

a. 474 ______

b. **7**0,158 ______

c. 18**7**,943 ______

d. 2,**7**31,008 ______

SRB 4

4. Add.

a. $3 + 5 =$ ______

b. $30 + 50 =$ ______

c. $300 + 500 =$ ______

d. ______ $= 9 + 7$

e. ______ $= 90 + 70$

f. ______ $= 900 + 700$

5. Solve the riddle.

I am a 2-dimensional figure.
I have two pairs of parallel sides.
All my angles have the same measure.
All my sides are the same length.

What am I? ______

SRB 86

6. Circle the pair of concentric circles.

Name-Collection Boxes

Write 5 names in each box below. Use as many different kinds of numbers and operations as you can. Draw an asterisk (*) next to the name you find most interesting.

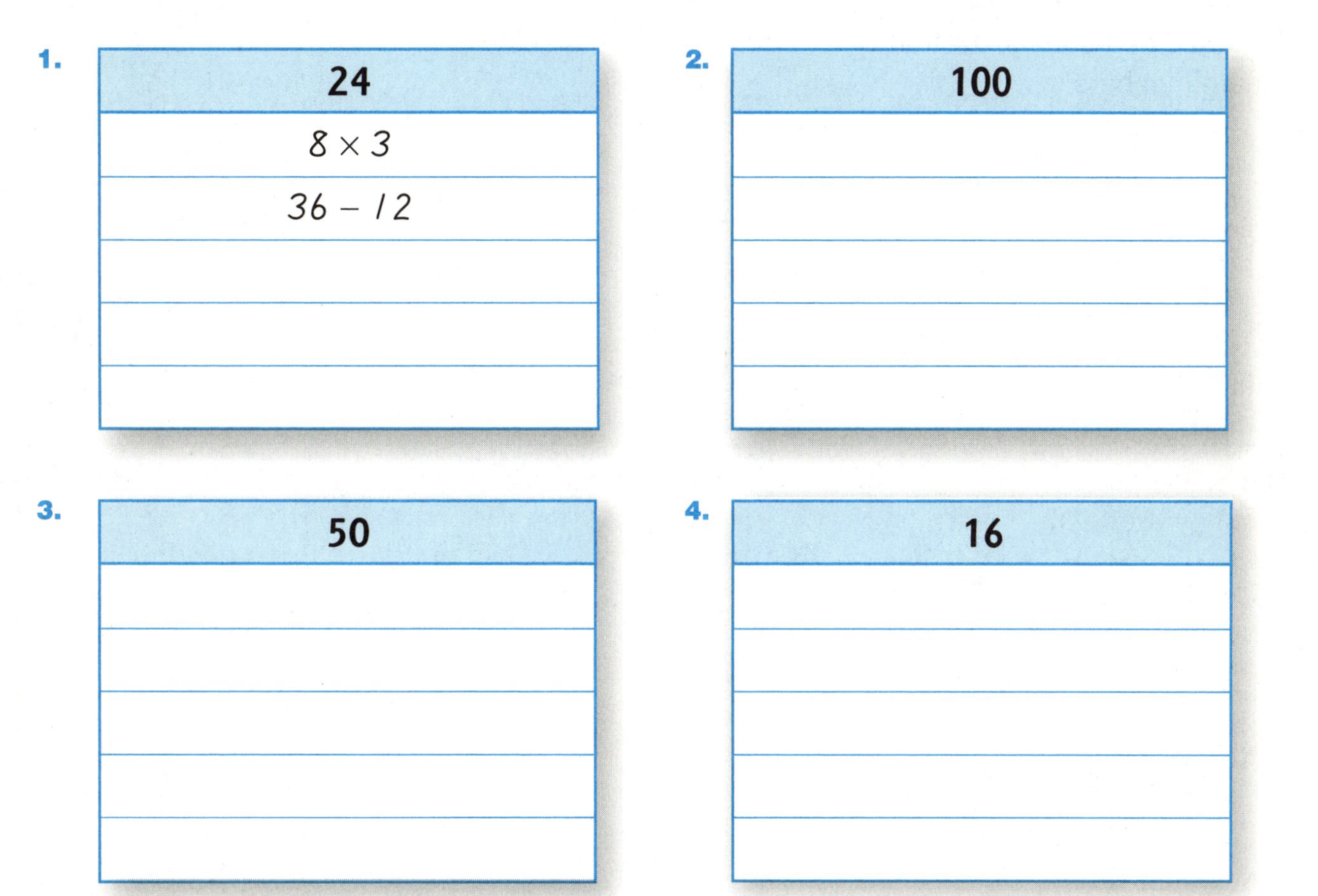

Write in your own numbers and fill in the boxes.

5.

6.

Date Time

Math Boxes 2.2

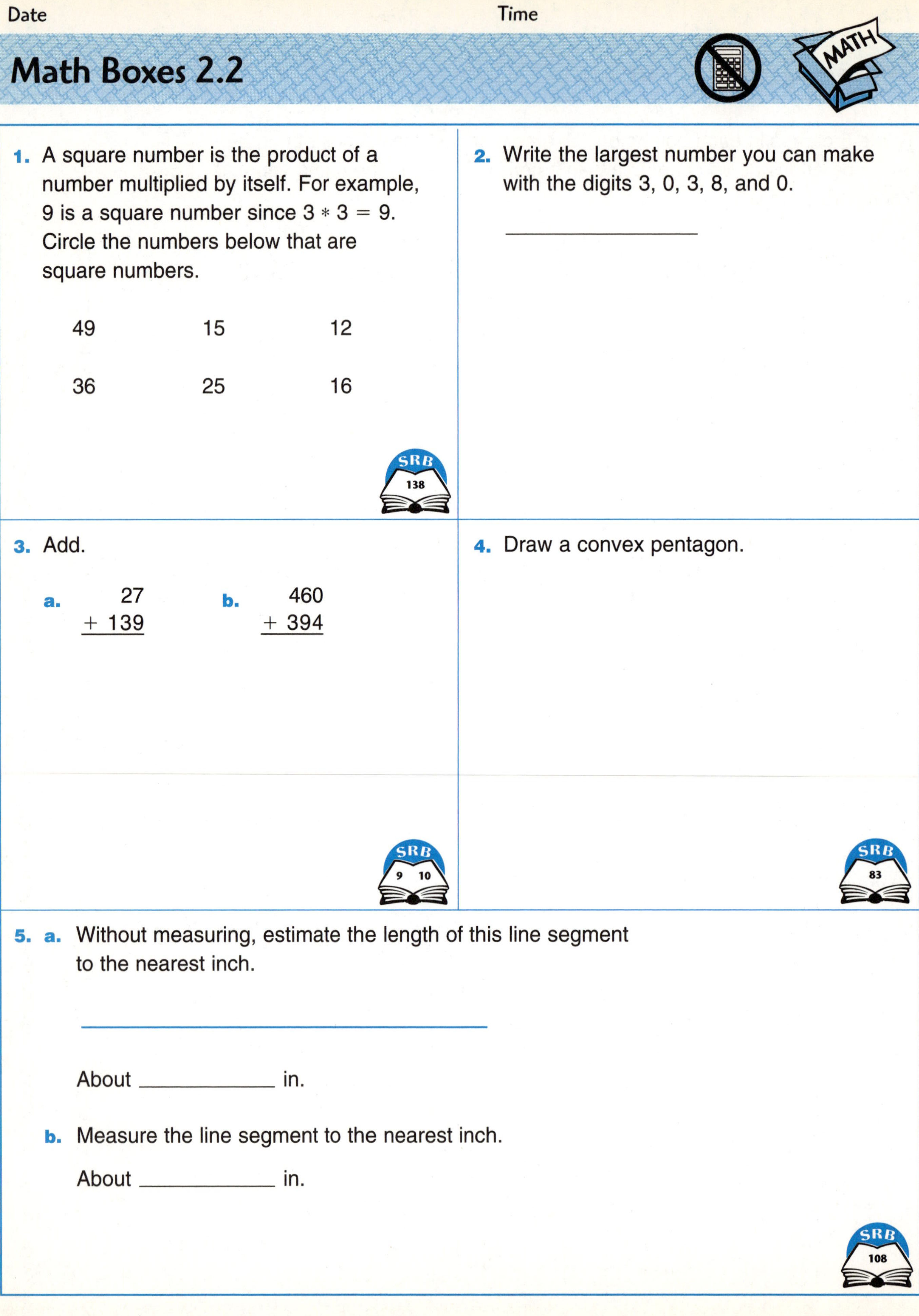

1. A square number is the product of a number multiplied by itself. For example, 9 is a square number since $3 * 3 = 9$. Circle the numbers below that are square numbers.

49	15	12
36	25	16

SRB 138

2. Write the largest number you can make with the digits 3, 0, 3, 8, and 0.

____

3. Add.

a. $27 + 139$

b. $460 + 394$

SRB 9 10

4. Draw a convex pentagon.

SRB 83

5. **a.** Without measuring, estimate the length of this line segment to the nearest inch.

About __________ in.

b. Measure the line segment to the nearest inch.

About __________ in.

SRB 108

Date Time

Place-Value Chart

Number	Hundred-Millions	Ten-Millions	Millions	Hundred-Thousands	Ten-Thousands	Thousands	Hundreds	Tens	Ones
	100M	10M	M	100K	10K	K	H	T	O

Use with Lesson 2.3.

Date ______ Time ______

Taking Apart, Putting Together

Complete.

1. In 574,

   5 is worth ___500___

   7 is worth ______

   4 is worth ______

2. In 9,027,

   9 is worth ______

   0 is worth ______

   2 is worth ______

   7 is worth ______

3. In 280,743,

   8 is worth ______

   2 is worth ______

   4 is worth ______

4. In 56,010,837,

   6 is worth ______

   1 is worth ______

   5 is worth ______

5. In 705,622,463,

   5 is worth ______

   4 is worth ______

   7 is worth ______

6. In 123,456,789,

   4 is worth ______

   9 is worth ______

   2 is worth ______

Add.

7.
```
   900
    70
+    5
------
```

8.
```
 30,000
  7,000
     50
+     2
-------
```

9.
```
 50,000,000
  9,000,000
     60,000
      2,000
        800
+        50
-----------
```

10.
```
 300,000,000
   9,000,000
     200,000
      70,000
          30
+          1
------------
```

Date Time

Math Boxes 2.3

1. Complete the division facts.

   a. $63 \div 9 =$ ______

   b. $36 \div$ ______ $= 9$

   c. $48 \div$ ______ $= 6$

   d. $60 \div$ ______ $= 6$

   e. $30 \div 6 =$ ______

   f. $81 \div 9 =$ ______

2. A sailfish can swim at a speed of 110 kilometers per hour. A tiger shark can swim at a speed of 53 kilometers per hour. How much faster can a sailfish swim than a tiger shark?

   ______ kilometers per hour

3. What is the value of the digit 8 in the numerals below?

   a. 5**8**4 ______

   b. 3**8**,067 ______

   c. 49,**8**41 ______

   d. **8**20,731 ______

4. Add.

   a. $4 + 5 =$ ______

   b. $40 + 50 =$ ______

   c. $400 + 500 =$ ______

   d. ______ $= 5 + 8$

   e. ______ $= 50 + 80$

   f. ______ $= 500 + 800$

5. Solve the riddle.

   I am a 2-dimensional figure.
   I have two pairs of parallel sides.
   None of my angles is a right angle.
   All my sides are the same length.

   What am I? ______

6. Use your compass to draw a pair of concentric circles.

Date Time

Math Boxes 2.4

1. Circle the numbers below that are square numbers.

   34 4 64

   81 88 100

2. Write the largest number you can make with the digits 5, 2, 3, 0, 6, 0, and 4.

   ____

3. Add.

   a. $\begin{array}{r} 145 \\ +\ 36 \\ \hline \end{array}$

   b. $\begin{array}{r} 290 \\ +\ 136 \\ \hline \end{array}$

4. Draw a concave pentagon.

5. a. Without measuring, estimate the length of this line segment to the nearest inch.

   About __________ in.

   b. Measure the line segment to the nearest inch.

   About __________ in.

Date Time

Counting Raisins

1. Use your $\frac{1}{2}$-ounce box of raisins. Complete each step when the teacher tells you. Stop after you complete each step.

   a. Don't open your box yet. **Guess** about how many raisins are in the box.

   About ________ raisins

   b. Open the box. Count the number of raisins in the top layer. Then **estimate** the total number of raisins in the box.

   About ________ raisins

   c. Now **count** the raisins in the box.

   How many? ________ raisins

2. Make a tally chart of the class data.

Number of Raisins	Number of Boxes

3. Find the following **landmarks** for the class data.

   a. What is the **maximum,** or largest, number of raisins found? ________

   b. What is the **minimum,** or smallest, number of raisins found? ________

   c. What is the **range**? (Subtract the minimum from the maximum.) ________

   d. What is the **mode,** or most frequent number of raisins found? ________

Date ______ Time ______

Math Boxes 2.5

1. Complete the multiplication facts.

   a. 5 * 7 = ______

   b. 3 * ______ = 18

   c. 7 * ______ = 56

   d. 9 * ______ = 45

   e. 8 * 4 = ______

   f. 8 * 9 = ______

2. Name the two pairs of parallel sides in parallelogram *HIJK*.

   ______ and ______

   ______ and ______

   H I K J

   SRB 80

3. A number has

   6 in the hundreds place,
   1 in the millions place,
   2 in the tens place,
   8 in the hundred-thousands place,
   5 in the ones place,
   3 in the thousands place, and
   4 in the ten-thousands place.

   Write the number.

   ___, ___ ___ ___, ___ ___ ___

   SRB 4

4. Write 5 names for 34.

   ______

   ______

   ______

   ______

   ______

   SRB 131

5. a. Without measuring, estimate the length of this line segment to the nearest centimeter.

   About ______ cm

   b. Measure the line segment to the nearest centimeter.

   About ______ cm

   SRB 108

Date Time

Family Size

Follow your teacher's directions and complete each step.

1. How many people are in your family? ________
Write the number on a stick-on note.

2. Make a line plot of the family-size data for the class.
Use **X**s in place of stick-on notes.

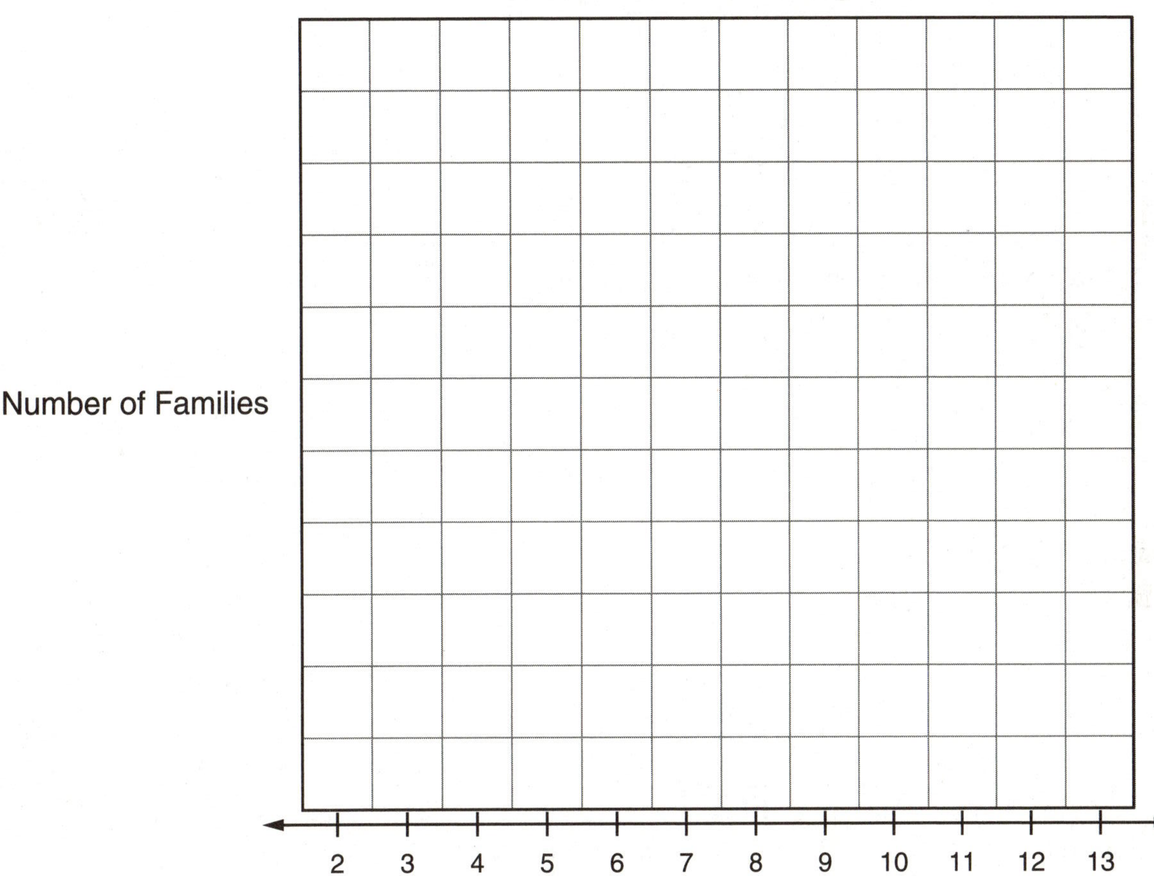

3. Find the following landmarks for the class data:

a. What is the **maximum** (largest) number of people in a family? ________

b. What is the **minimum** (smallest) number of people in a family? ________

c. What is the **range**? (Subtract the minimum from the maximum.) ________

d. What is the **mode** (most frequent family size)? ________

4. What is the **median** family size for the class? ________ people

Date Time

Math Boxes 2.6

1. Tell whether each number sentence is true or false.

a. $14 + 7 = 22$ ______

b. $45 - 12 = 33$ ______

c. $27 = 40 - 13$ ______

d. $36 = 15 + 15$ ______

SRB 128

2. A royal python can be 35 feet long. An anaconda can be 28 feet long. What would be their combined length, end-to-end?

______ feet

SRB 132-152

3. Find the following landmarks for this set of numbers: 3, 45, 15, 13, 3, 7, 19.

a. median ______

b. mode ______

c. maximum ______

d. minimum ______

e. range ______

SRB 65 66

4. Add.

a. $2 + 7 =$ ______

b. $20 + 70 =$ ______

c. $200 + 700 =$ ______

d. ______ $= 8 + 4$

e. ______ $= 80 + 40$

f. ______ $= 800 + 400$

5. Subtract.

a. $147 - 56$

b. $531 - 246$

SRB 11-14

6. Write 4,007,392 in words.

______

______

______

______

______

SRB 4

Date Time

Partial-Sums Addition

Solve Problems 1–3 using the partial-sums method. Solve Problems 4–6 using any method you choose. Show your work.

1. 76 + 38 = ____________
2. 166 + 28 = ____________
3. ____________ = 647 + 936
4. ____________ = 6,236 + 645
5. ____________ = 1,672 + 3,221
6. 17,854 + 24,550 = ____________

Use with Lesson 2.7.

Date Time

Column Addition

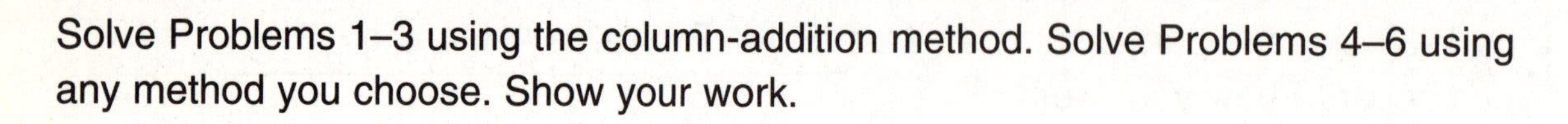

Solve Problems 1–3 using the column-addition method. Solve Problems 4–6 using any method you choose. Show your work.

1. 94 + 47 = ________

2. 549 + 33 = ________

3. ________ = 385 + 726

4. ________ = 9,046 + 946

5. ________ = 2,538 + 4,179

6. 24,614 + 73,058 = ________

Use with Lesson 2.7.

Date Time

Math Boxes 2.7

1. Complete the multiplication facts.

   a. 5 * 8 = ________

   b. 2 * ________ = 16

   c. 7 * ________ = 21

   d. 9 * ________ = 54

   e. 8 * 3 = ________

   f. 9 * 8 = ________

2. Draw a parallelogram. Label the vertices so that side *AB* is parallel to side *CD*.

3. A number has

   3 in the millions place,
   1 in the ones place,
   8 in the thousands place,
   9 in the ten-thousands place,
   0 in the tens place,
   6 in the hundred-thousands place, and
   5 in the hundreds place.

   Write the number.

   __, __ __ __, __ __ __

4. Write 5 names for 100.

   ____

   ____

   ____

   ____

   ____

5. a. Without measuring, estimate the length of this line segment to the nearest centimeter.

   About ________ cm

   b. Measure the line segment to the nearest centimeter.

   About ________ cm

Head Sizes

Ms. Woods owns a clothing store. She is trying to decide how many children's hats to stock in each possible size. Should she stock the same number of hats in each size? Or should she stock more hats in the more popular sizes?

Help Ms. Woods decide. Pretend that she has asked each class in your school to collect and organize data about students' head sizes. She plans to combine the data for all the classes and then use the data to figure out how many hats of each size to stock.

As a class, collect and organize data about one another's head sizes.

1. Ask your partner to help you measure the distance around your head.

- Wrap the tape measure once around your head.
- See where the tape touches the end tip of the tape measure.
- Read the mark where the tape touches the end tip.
- Read this length to the nearest $\frac{1}{2}$ centimeter.

Record your head size. About _______ cm

2. What is the median head size for the class? About _______ cm

What's Your Hat Size?

Men's and women's hats are sized in different ways. The size of a woman's hat—22, for example—is approximately the distance around the head in inches.

To understand a man's hat size—$6\frac{7}{8}$, for example—imagine taking the sweatband inside the hat and forming it into a circle. The size is the diameter of the circle to the nearest $\frac{1}{8}$ inch.

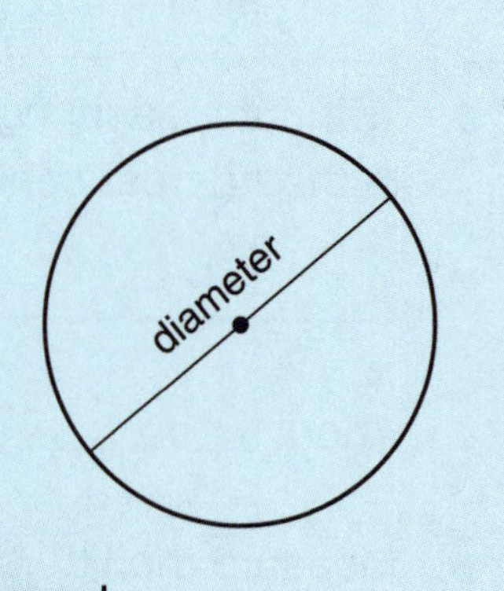

This system began centuries ago, when new hats were round. Men used a device called a hat screw to force their hats into a more comfortable oval shape.

Date Time

Head Sizes (cont.)

3. Make a bar graph of the head-size data for the class.

title

label

label

Date Time

Addition of Multidigit Numbers

Use your favorite method to solve the problems.

1. 78 + 495 = ____________

2. ____________ = 234 + 681

3. 907 + 388 = ____________

4. 6,891 + 736 = ____________

5. ____________ = 3,801 + 9,777

6. ____________ = 23,408 + 67,993

Date ______ Time ______

Math Boxes 2.8

1. Tell whether each number sentence is true or false.

   a. $18 + 9 = 37$ ______

   b. $29 = 17 + 12$ ______

   c. $17 = 40 - 24$ ______

   d. $42 - 15 = 27$ ______

2. An ostrich can weigh about 345 pounds. An emu can weigh about 88 pounds. How much do they weigh altogether?

   ______ pounds

3. Find the following landmarks for this set of numbers: 12, 16, 23, 15, 16, 19, 18.

   a. median ______

   b. mode ______

   c. maximum ______

   d. minimum ______

   e. range ______

4. Add.

   a. $2 + 4 =$ ______

   b. $20 + 40 =$ ______

   c. $200 + 400 =$ ______

   d. ______ $= 8 + 6$

   e. ______ $= 80 + 60$

   f. ______ $= 800 + 600$

5. Subtract.

   a. $\begin{array}{r} 231 \\ -\ 84 \\ \hline \end{array}$

   b. $\begin{array}{r} 603 \\ -\ 466 \\ \hline \end{array}$

6. Write 8,042,176 in words.

   ______

   ______

   ______

   ______

   ______

Date Time

Trade-First Subtraction

Solve Problems 1–3 using the trade-first method. Solve Problems 4–6 using any method you choose. Show your work.

1. 58 – 32 = ____________
2. 73 – 37 = ____________
3. ____________ = 652 – 379
4. ____________ = 900 – 461
5. ____________ = 2,936 – 1,652
6. 2,468 – 789 = ____________

Date Time

Partial-Differences Subtraction

Solve Problems 1–3 using the partial-differences method. Solve Problems 4–6 using any method you choose. Show your work.

1. 86 − 34 = ____________

2. 61 − 26 = ____________

3. ____________ = 942 − 485

4. ____________ = 400 − 271

5. ____________ = 7,584 − 2,806

6. 1,681 − 893 = ____________

Date Time

Subtraction by Counting Up

Solve the problems in your head. Use the counting-up strategy.

Example

50 − 26 = ?

Think: 26 + **4** = 30

30 + **20** = 50

4 + 20 = 24

So, 50 − 26 = 24

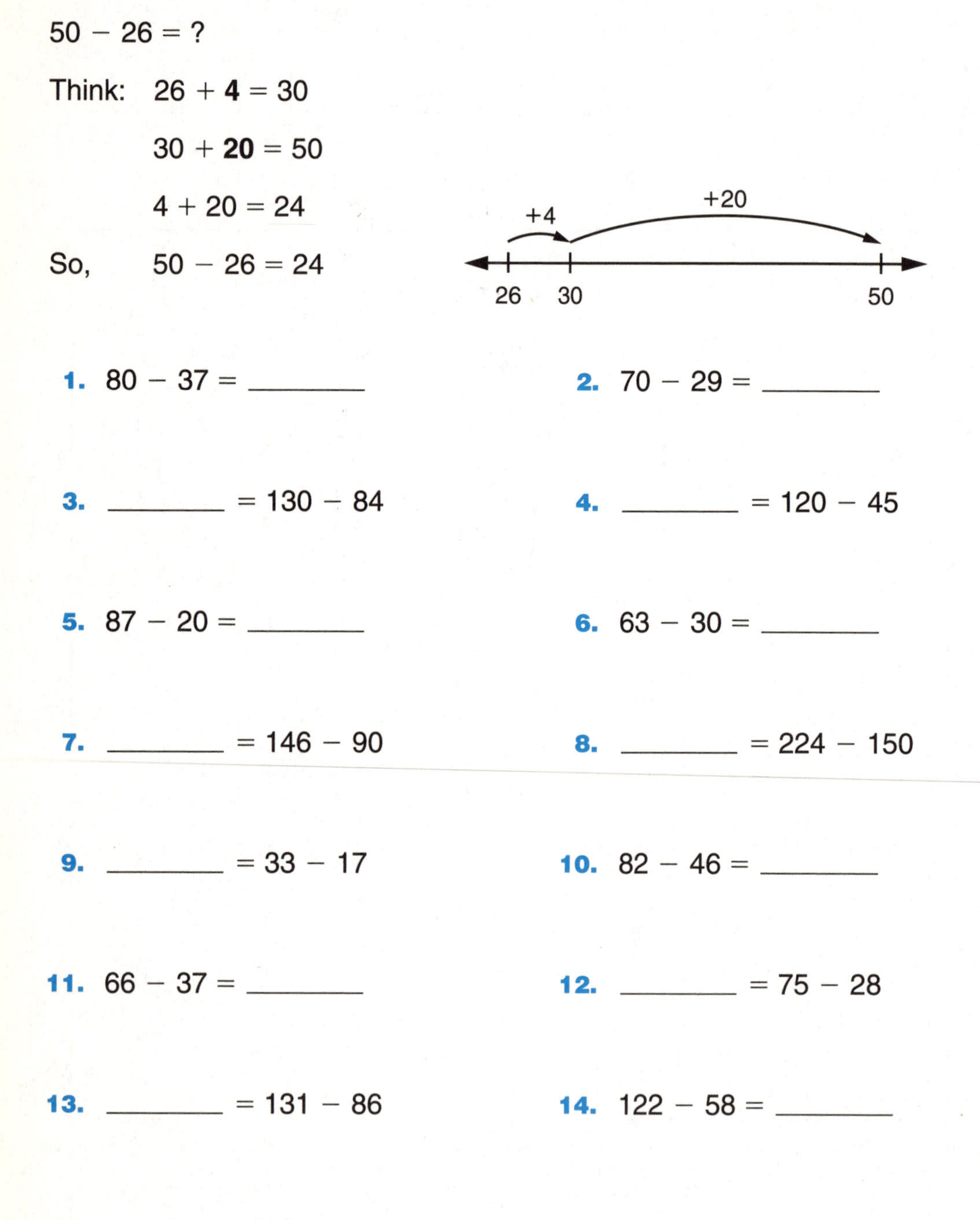

1. 80 − 37 = ________

2. 70 − 29 = ________

3. ________ = 130 − 84

4. ________ = 120 − 45

5. 87 − 20 = ________

6. 63 − 30 = ________

7. ________ = 146 − 90

8. ________ = 224 − 150

9. ________ = 33 − 17

10. 82 − 46 = ________

11. 66 − 37 = ________

12. ________ = 75 − 28

13. ________ = 131 − 86

14. 122 − 58 = ________

Date Time

Math Boxes 2.9

1. Complete the multiplication facts.

a. 5 * 4 = ______

b. 7 * ______ = 63

c. 4 * ______ = 24

d. 6 * ______ = 48

e. 9 * 3 = ______

f. 5 * 6 = ______

2. Draw a shape that has no parallel sides.

3. A number has

6 in the tens place,
9 in the millions place,
4 in the thousands place,
3 in the ten-thousands place,
2 in the hundred-thousands place,
0 in the ones place, and
8 in the hundreds place.

Write the number.

___, ___ ___ ___, ___ ___ ___

4. Write 5 names for 1,000.

______

______

______

______

______

5. a. Without measuring, estimate the length of this line segment to the nearest centimeter.

About ______ cm

b. Measure the line segment to the nearest centimeter.

About ______ cm

Date Time

Time to Reflect

1. As you discovered the many uses of numbers, did any of the uses surprise you? Why? Why not?

2. When you are filling in a name-collection box, do you find that you tend to use one operation more than any other? How could you break this habit? If you do not have this habit, what advice would you give a classmate to help him or her think creatively about different names for numbers?

3. What is the hardest part of adding or subtracting multidigit numbers? What did you do to overcome these difficulties?

Date Time

Math Boxes 2.10

1. Complete the division facts.

   a. 35 ÷ 5 = ______

   b. 56 ÷ ______ = 8

   c. 32 ÷ ______ = 4

   d. 24 ÷ ______ = 6

   e. 72 ÷ 8 = ______

   f. 40 ÷ 5 = ______

2. Complete the multiplication facts.

   a. 4 * 7 = ______

   b. 3 * ______ = 15

   c. 7 * ______ = 42

   d. 9 * ______ = 36

   e. 6 * 0 = ______

   f. 1 * 9 = ______

3. Complete the square facts.

   a. 64 ÷ 8 = ______

   b. 49 ÷ 7 = ______

   c. 4 * 4 = ______

   d. 3 * 3 = ______

   e. 25 ÷ 5 = ______

4. Tell whether each number sentence is true or false.

   a. 46 + 12 = 53 ______

   b. 36 = 22 + 14 ______

   c. 13 = 84 − 71 ______

   d. 52 − 20 = 34 ______

5. A grizzly bear can weigh 786 pounds. An American black bear can weigh 227 pounds. What is their combined weight?

   ______ pounds

6. On the average, India produces about 851 movies per year. The United States produces about 569 movies. On the average, how many fewer movies per year does the United States produce than India?

   ______

Date Time

Multiplication/Division Facts Table

*, /	1	2	3	4	5	6	7	8	9	10
1	1	2	3	4	5	6	7	8	9	10
2	2	4	6	8	10	12	14	16	18	20
3	3	6	9	12	15	18	21	24	27	30
4	4	8	12	16	20	24	28	32	36	40
5	5	10	15	20	25	30	35	40	45	50
6	6	12	18	24	30	36	42	48	54	60
7	7	14	21	28	35	42	49	56	63	70
8	8	16	24	32	40	48	56	64	72	80
9	9	18	27	36	45	54	63	72	81	90
10	10	20	30	40	50	60	70	80	90	100

Date Time

Math Boxes 3.1

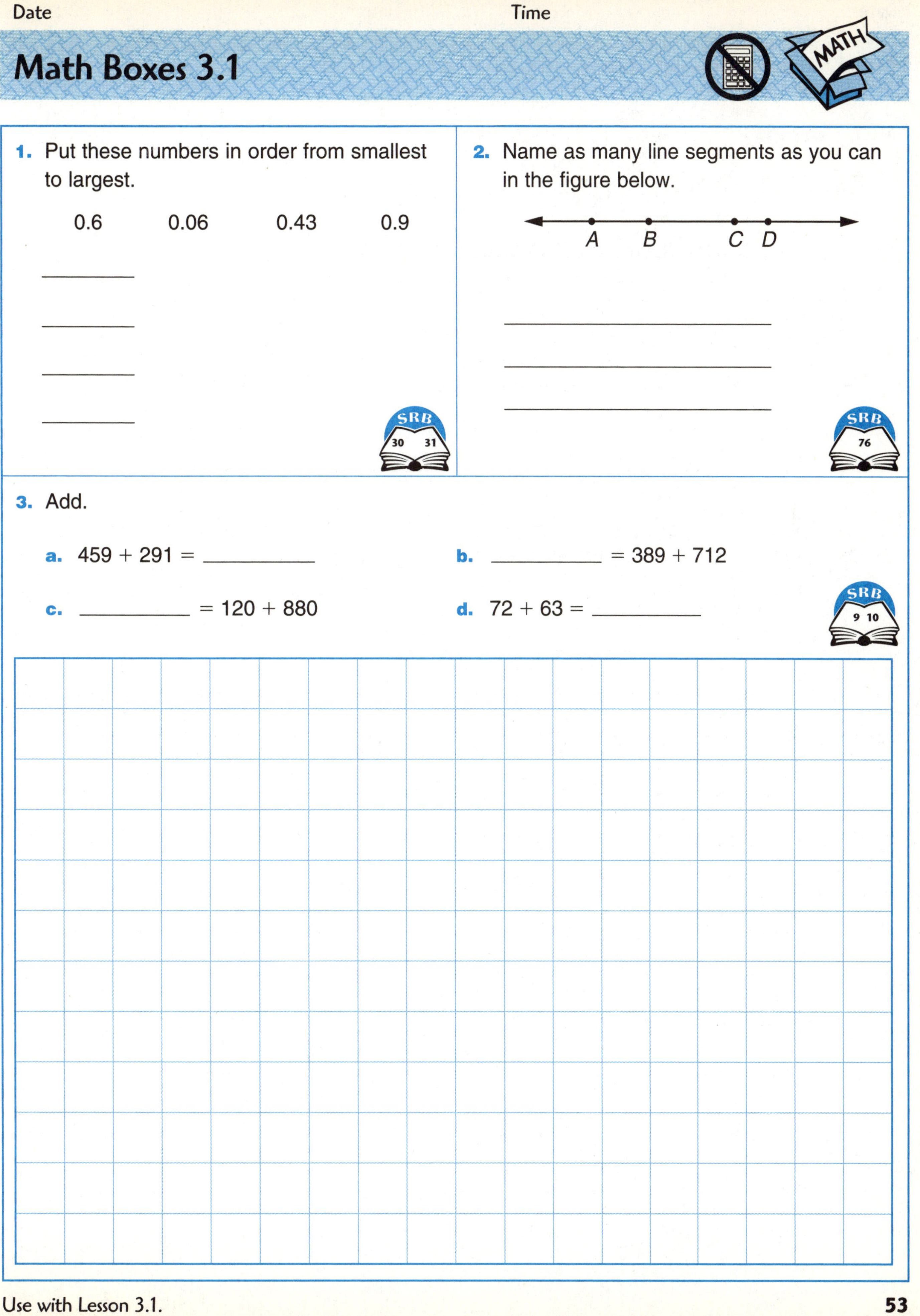

1. Put these numbers in order from smallest to largest.

0.6 0.06 0.43 0.9

______

______

______

______

SRB 30 31

2. Name as many line segments as you can in the figure below.

______

______

______

SRB 76

3. Add.

a. 459 + 291 = ______

b. ______ = 389 + 712

c. ______ = 120 + 880

d. 72 + 63 = ______

SRB 9 10

Date Time

Patterns in Multiplication Facts

Math Message

Look at the table on journal page 52.

1. Find a pattern in the 9s multiplication facts. Describe the pattern.

2. Find a pattern in the 5s multiplication facts. Describe the pattern.

3. What other patterns can you find in the multiplication facts? Write about some of them.

Date Time

Math Boxes 3.2

1. Which amount below is closest to the sum of $2.50, $0.75, $3.85, and $12.70? Circle the amount.

   $5.00

   $10.00

   $15.00

   $20.00

SRB 155

2. Number of days it took 10 students to complete their science projects:

   6, 4, 10, 11, 8, 6, 14, 9, 3, 12

   a. What is the range for this set of numbers? ________

   b. What is the median? ________

SRB 65

3. Subtract.

   a. ________ = 81 − 46

   b. ________ = 372 − 84

   c. 602 − 213 = ________

   d. 420 − 167 = ________

SRB 11-14

Date Time

Math Boxes 3.3

1. Put these numbers in order from smallest to largest.

0.7 0.007 0.5 0.63

____

____

____

____

2. Name as many line segments as you can in the figure below.

E F G H

____

____

____

3. Add.

a. $379 + 685 =$ __________

b. __________ $= 208 + 397$

c. __________ $= 230 + 570$

d. $34 + 76 =$ __________

Date ____________ Time ____________

Multiplication and Division

Equivalents			
$3 * 4$	$12 / 3$	$12 \div 3$	$3 < 5$ ($<$ means "is less than")
3×4	$\frac{12}{3}$	$3\overline{)12}$	$5 > 3$ ($>$ means "is greater than")

1. Choose 3 Fact Triangles. Write the fact family for each.

____ * ____ = ____	____ × ____ = ____	____ * ____ = ____
____ * ____ = ____	____ × ____ = ____	____ * ____ = ____
____ / ____ = ____	____ / ____ = ____	____ ÷ ____ = ____
____ / ____ = ____	____ / ____ = ____	____ ÷ ____ = ____

2. Solve each division fact.

a. $27 / 3 =$ ____ Think: How many 3s in 27?

b. ____ $= 45 / 5$ Think: How many 5s in 45?

c. $36 \div 6 =$ ____ Think: 6 times what number equals 36?

d. $24 / 8 =$ ____ Think: 8 times what number equals 24?

3. A cashier has 5 rolls of quarters and 6 rolls of dimes in his cash register. Each roll of quarters is worth $10, and each roll of dimes is worth $5.

a. How much are the rolls of quarters and dimes worth in all? $________

b. How many quarters are in 1 roll? ________ quarters

c. How many quarters are in the 5 rolls? ________ quarters

d. How many dimes are in 1 roll? ________ dimes

e. How many dimes are in the 6 rolls? ________ dimes

f. There is also $7.50 worth of half-dollars in the cash register. How many half-dollars is that? ________ half-dollars

Date Time

Math Boxes 3.4

1. Which amount below is closest to the sum of $1.80, $3.75, $1.85, and $1.70? Circle the amount.

   $5.00

   $10.00

   $15.00

   $20.00

2. Number of glasses of milk drunk by 10 students in a week:

   16, 13, 15, 20, 8, 10, 15, 12, 10, 18

   a. What is the range for this set of numbers? ________

   b. What is the median? ________

3. Subtract.

   a. ________ = 72 − 35

   b. ________ = 489 − 95

   c. 503 − 121 = ________

   d. 830 − 342 = ________

Date Time

Math Boxes 3.5

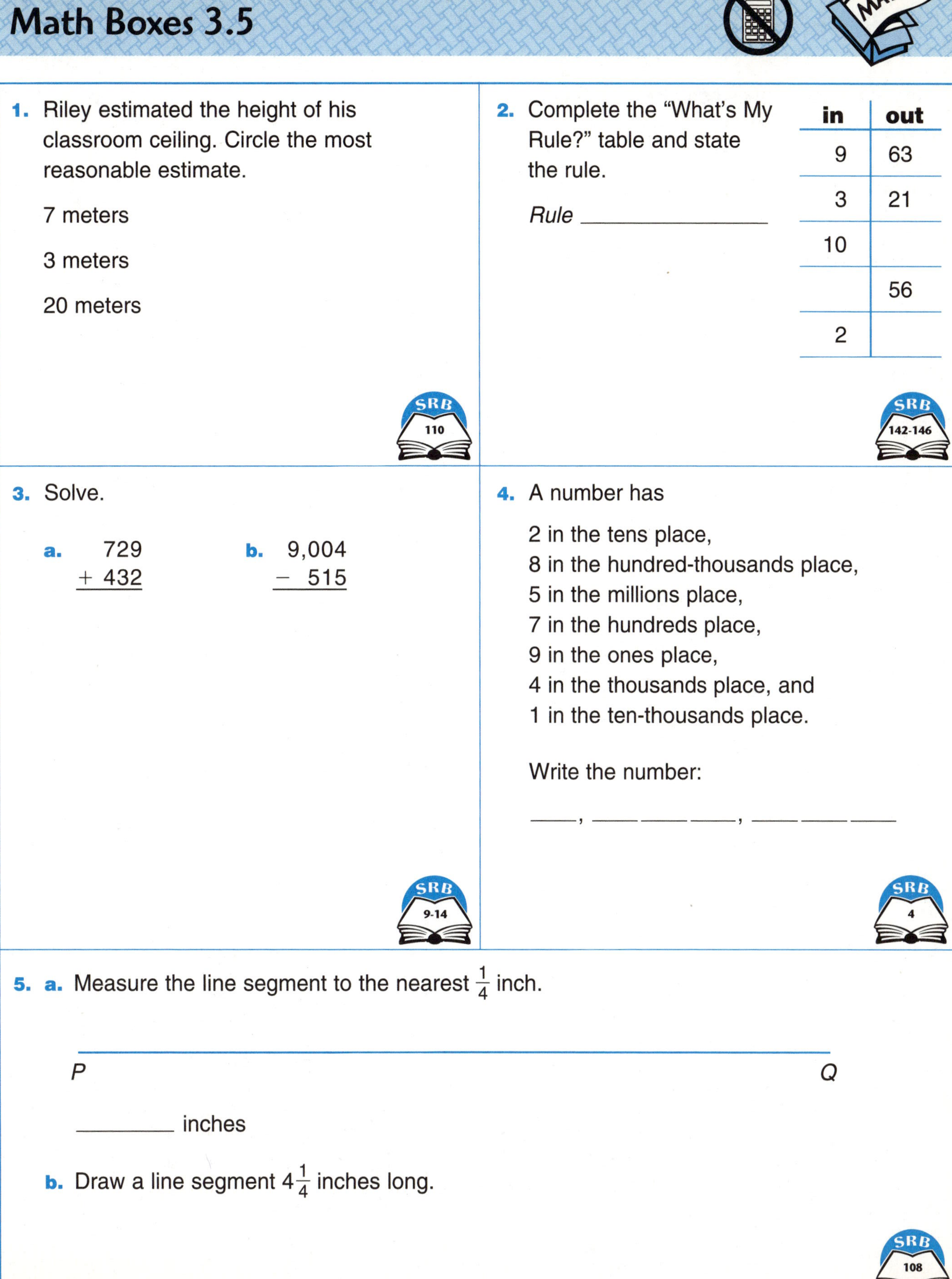

1. Riley estimated the height of his classroom ceiling. Circle the most reasonable estimate.

7 meters

3 meters

20 meters

SRB 110

2. Complete the "What's My Rule?" table and state the rule.

Rule ______________

in	out
9	63
3	21
10	
	56
2	

SRB 142-146

3. Solve.

a.
$$\begin{array}{r} 729 \\ +\ 432 \\ \hline \end{array}$$

b.
$$\begin{array}{r} 9{,}004 \\ -\ 515 \\ \hline \end{array}$$

SRB 9-14

4. A number has

2 in the tens place,
8 in the hundred-thousands place,
5 in the millions place,
7 in the hundreds place,
9 in the ones place,
4 in the thousands place, and
1 in the ten-thousands place.

Write the number:

___ , ___ ___ ___ , ___ ___ ___

SRB 4

5. a. Measure the line segment to the nearest $\frac{1}{4}$ inch.

P ________________________________ Q

________ inches

b. Draw a line segment $4\frac{1}{4}$ inches long.

SRB 108

Date Time

Measuring Air Distances

Estimate which city listed below is the closest to Washington, D.C. ____________

Estimate which city is the farthest. ____________

Air distance from Washington, D.C. to:

Cairo, Egypt About ____________ miles

Mexico City, Mexico About ____________ miles

Stockholm, Sweden About ____________ miles

Moscow, Russia About ____________ miles

Tokyo, Japan About ____________ miles

Shanghai, China About ____________ miles

Sydney, Australia About ____________ miles

Warsaw, Poland About ____________ miles

Cape Town, South Africa About ____________ miles

Rio de Janeiro, Brazil About ____________ miles

Choose your own city. ____________ About ____________ miles

Earth

The distance around the equator is about 42 million times the distance around a globe with a 12-inch diameter.

If Earth were a huge, empty ball, it would be possible to fill it with about 50,000,000,000,000,000,000,000 twelve-inch globes. This number is read as "50 sextillion." Another way to write this number is $5 * 10^{22}$. This is read as "5 times 10 to the 22nd power."

Date Time

Math Boxes 3.6

1. What numbers come next?

a. 0.1, 0.2, 0.3, _____, _____, _____

b. 1.4, 1.6, 1.8, _____, _____, _____

c. 3, 2.5, 2, _____, _____, _____

d. 6.72, 6.73, 6.74, _______, _______, _______

SRB 34 35

2. Add or subtract.

a. $3.56 + $2.49 = __________

b. $0.67 + $0.08 = __________

c. $6.25 − $5.01 = __________

d. $3.37 − $0.24 = __________

SRB 34 35

3. Complete the "What's My Rule?" table and state the rule.

Rule ________________

in	out
40	5
24	3
64	
	4
	9

SRB 145

4. If 1 centimeter on a map represents 20 kilometers, then

a. 2 cm represent ______ km.

b. 8 cm represent ______ km.

c. 5 mm represent ______ km.

d. 3.5 cm represent ______ km.

e. 6.5 cm represent ______ km.

SRB 125

5. A rock collector has 136 rocks in her collection. She took them to a geologist who said that 57 of them are not valuable. How many of them are valuable?

____

SRB 132 152

6. Solve the riddle.

I am a four-sided polygon.

My two short sides are the same length and meet at a vertex.

My two long sides are the same length and meet at a vertex.

What am I? __________

SRB 86

Date Time

Flying to Cairo

Pretend that you are flying from Washington, D.C., to Cairo, Egypt. You have a choice of flying by way of Amsterdam, Holland, or by way of Rome, Italy. The air distances are shown in the table.

Air Distance between Cities (in miles)

	Amsterdam	Rome
Washington, D.C.	3,851	4,497
Cairo	2,035	1,326

If you fly first class, your ticket will cost $850.
If you fly economy class, you will save $175.

1. About how many more miles is it from Washington, D.C., to Rome than from Washington, D.C., to Amsterdam?

   ____
   (number model)

   Answer: About ________ miles

2. What is the total distance from Washington, D.C., to Cairo by way of Amsterdam?

   ____
   (number model)

   Answer: About ________ miles

3. What is the total distance from Washington, D.C., to Cairo by way of Rome?

   ____
   (number model)

   Answer: About ________ miles

Date Time

Flying to Cairo (cont.)

4. About how many fewer miles will you fly if you go by way of Rome rather than Amsterdam?

____
(number model)

Answer: About __________ miles

5. How much will your ticket cost if you fly economy class?

____
(number model)

Answer: __________

6. What is the total distance (round-trip) from Washington, D.C., to Cairo by way of Rome, and then from Cairo back to Washington, D.C., by way of Rome again?

____
(number model)

Answer: About __________ miles

7. One of the flights to Amsterdam leaves Washington, D.C., at 1:15 P.M. It lands in Amsterdam 8 hours and 45 minutes later.

a. At what time does it land, Washington, D.C., time?

Answer: ______________ (A.M. or P.M.?)

b. At what time does it land, Amsterdam time? (Use the Time Zones map on pages 220 and 221 in the World Tour section of your *Student Reference Book.*)

Answer: ______________ (A.M. or P.M.?)

Date Time

Math Boxes 3.7

1. Sara measured the length of her arm. Circle the most reasonable measurement.

5 centimeters

50 centimeters

150 centimeters

2. Complete the "What's My Rule?" table and state the rule.

Rule ________________

in	out
2	8
6	24
	32
4	
	20

3. Solve.

a. $604 + 817$

b. $3{,}005 - 686$

4. A number has

1 in the tens place,
3 in the hundreds place,
5 in the ones place,
7 in the hundred-thousands place,
9 in the thousands place,
2 in the ten-thousands place, and
4 in the millions place.

Write the number:

___, ___ ___ ___, ___ ___ ___

5. a. Measure the line segment to the nearest $\frac{1}{4}$ inch.

B ______________________________ *L*

________ inches

b. Draw a line segment $4\frac{3}{4}$ inches long.

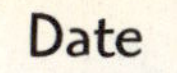

Number Sentences

Tell whether each sentence below is true or false. If it is not possible to tell, write "can't tell" on the answer blank.

1. $28 - 16 = 12$ ______
2. $7 + 3 < 1$ ______
3. $6 = 36 \div 6$ ______
4. $27 + 3 = 5 * 6$ ______
5. $80 - ? = 40$ ______
6. $0 = 4 / 4$ ______
7. $3 * 8 < 30$ ______
8. 2×7 ______

9. Make up 5 true number sentences and 5 false number sentences. Mix them up. Ask your partner to tell whether each sentence is true or false.

Equivalents			
$3 * 4$	$12 / 3$	$12 \div 3$	$3 < 5$ (< means "is less than")
3×4	$\frac{12}{3}$	$3\overline{)12}$	$5 > 3$ (> means "is greater than")

a. ______

b. ______

c. ______

d. ______

e. ______

f. ______

g. ______

h. ______

i. ______

j. ______

Date Time

Elapsed Time

During a solar eclipse, the moon passes between the sun and Earth and casts a shadow across the surface of Earth. When this happens, it gets darker, as if it were dusk, even if the eclipse occurs in the middle of the day.

Total solar eclipses, in which the moon blocks out the entire sun, are very rare. A near-total eclipse, in which about 90% of the sun is hidden from view, occurred over part of the United States on May 10, 1994. Before that, the most recent total eclipse visible in the continental United States occurred in 1979.

1. How many years had passed between these two eclipses? ______________ years

The next total solar eclipse will not be seen over the United States until 2017.

2. How old will you be then? ______________ years old

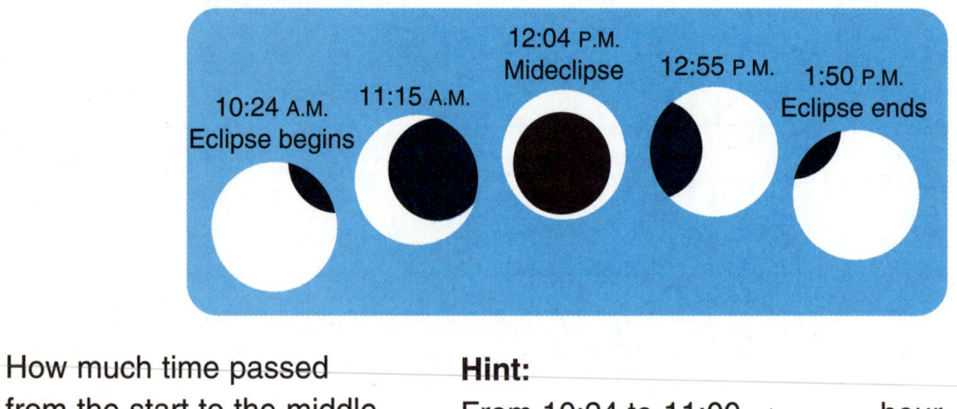

3. How much time passed from the start to the middle of the eclipse?

____

Hint:

From 10:24 to 11:00 → _____ hour _____ minutes

From 11:00 to 12:04 → _____ hour _____ minutes

Total → _____ hour _____ minutes

4. How much time passed from the middle to the end of the eclipse? ______________________

5. How long did the eclipse last? ______________________

Eclipses

People in ancient times were afraid of solar eclipses. The Chinese thought the eclipse was caused by a dragon swallowing the sun. They would bang on pans and shoot arrows into the sky to scare the dragon. The Japanese believed that poison fell from the sky. They would cover their wells to prevent the poison from getting into their drinking water.

Date Time

Math Boxes 3.8

1. What numbers come next?

a. 0.5, 0.8, 1.1, ______, ______, ______

b. 2.7, 3.2, 3.7, ______, ______, ______

c. 4, 3.8, 3.6, ______, ______, ______

d. 5.05, 5.06, 5.07, ______, ______, ______

2. Add or subtract.

a. $10.97 + $15.60 = ______

b. $1.23 + $7.00 = ______

c. $4.56 – $2.07 = ______

d. $1.85 – $1.53 = ______

3. Complete the "What's My Rule?" table and state the rule.

Rule ______

in	out
45	5
81	9
	3
	4
72	

4. If 1 inch on a map represents 30 miles, then

a. 3 inches represent ______ miles.

b. 9 inches represent ______ miles.

c. $\frac{1}{2}$ inch represents ______ miles.

d. $5\frac{1}{2}$ inches represent ______ miles.

e. 1 foot represents ______ miles.

5. The statue of Chief Crazy Horse in South Dakota is 563 feet tall. The Statue of Liberty is 151 feet tall. What is the difference in height of the two statues?

______ feet

6. Solve the riddle.

I am a polygon.

All my angles have the same measure.

Each of my 5 sides has the same measure.

What am I? ______

Date Time

Parentheses in Number Sentences

Part 1

Make a true sentence by filling in the missing number.

1. a. $(30 - 15) * 2 =$ ________ b. $30 - (15 * 2) =$ ________

2. a. ________ $= 28 / (14 / 2)$ b. ________ $= (28 / 14) / 2$

3. a. $(6 + 8) / (2 - 1) =$ ________ b. $6 + (8/2) - 1 =$ ________

Part 2

Make a true sentence by inserting parentheses.

4. a. $4 \times 9 - 2 = 34$ b. $4 \times 9 - 2 = 28$

5. a. $24 = 53 - 11 + 18$ b. $60 = 53 - 11 + 18$

6. a. $12 / 4 + 2 = 2$ b. $12 / 4 + 2 = 5$

7. a. $55 = 15 + 10 \times 4$ b. $100 = 15 + 10 \times 4$

Challenge

8. a. $10 - 4 / 2 * 3 = 24$ b. $10 - 4 / 2 * 3 = 1$

Part 3

Pretend you are playing a game of *Name That Number* with only 3 cards per hand. To name the target number, use all 3 numbers and any operations you want. For each problem, write a true number sentence, containing parentheses, using the 3 numbers and the target number.

9. Use: 2, 5, 15 Target number: 5 ____________________

10. Use: 3, 4, 5 Target number: 17 ____________________

11. Use: 1, 3, 11 Target number: 4 ____________________

Date Time

Math Boxes 3.9

1. a. Measure the line segment to the nearest centimeter.

B ______________________________ N

_______ cm

b. Draw a line segment that is half the length of $\overline{BN}$.

c. How long is the line segment you drew? _______ cm

SRB 108

2. QUARTZY is the highest scoring word in a popular board game; it is worth 164 points. BEZIQUE is the second-highest scoring word, worth 161 points. If you were able to use both words, what would your score be?

_______ points

SRB 132 152

3. Complete the multiplication facts.

a. 4 * 5 = _______

b. 6 * 6 = _______

c. _______ = 8 * 3

d. _______ = 6 * 7

e. _______ = 5 * 5

f. 6 * 8 = _______

4. Tell whether each number sentence is true or false.

a. 55 = 10 + (5 * 7) _______

b. 74 = (4 * 7) + (7 * 6) _______

c. 68 > (9 * 4) + 13 _______

d. 4 = (8 * 7) / (12 + 2) _______

SRB 130

5. Make a true sentence by inserting parentheses.

a. 8 * 2 + 20 = 36

b. 63 = 9 / 3 * 21

c. 28 / 7 * 4 = 16

d. 10 / 5 * 2 = 1

SRB 129

Date Time

Broken Calculator

Solve each open sentence on your calculator without using the "broken" key.
Record your steps.

1.

Broken Key: (−)

To Solve: $68 + x = 413$

2.

Broken Key: (÷)

To Solve: $s * 48 = 2,928$

3.

Broken Key: (−)

To Solve: $z + 643 = 1,210$

4.

Broken Key: (×)

To Solve: $w / 15 = 8$

5.

Broken Key: (+)

To Solve: $d - 574 = 1,437$

6. Make up one for your partner to solve.

Broken Key:

To Solve:

Use with Lesson 3.10.

Date Time

Open Sentences

Solve each open sentence. Copy the sentence over, with the solution in place of the variable. Circle the solution.

1. $51 = n + 29$

   $51 = (22) + 29$

2. $48 + d = 70$

   ____

3. $34 - x = 7$

   ____

4. $32 = 76 - p$

   ____

5. $b - 7 = 12$

   ____

6. $u - 30 = 10$

   ____

7. $y = 3 * 8$

   ____

8. $5 * m = 35$

   ____

9. $21 / x = 7$

   ____

10. $x = 32 / 8$

    ____

11. $5 = w / 10$

    ____

12. $h - 6 = 9$

    ____

Date Time

Math Boxes 3.10

1. What numbers come next?

   a. 0.2, 0.6, 1, _____, _____, _____

   b. 5.6, 5.9, 6.2, _____, _____, _____

   c. 8, 7.4, 6.8, _____, _____, _____

   d. 7.03, 7.08, 7.13, _____, _____, _____

2. Add or subtract.

   a. $14.99 + $2.80 = _____

   b. $1.46 + $8.12 = _____

   c. $5.87 – $2.15 = _____

   d. $1.74 – $1.23 = _____

3. Complete the "What's My Rule?" table and state the rule.

   Rule _____

in	out
24	8
9	3
	6
	7
15	

4. If $\frac{1}{2}$ inch on a map represents 5 miles, then

   a. 1 in. represents _____ mi.

   b. $4\frac{1}{2}$ in. represent _____ mi.

   c. _____ in. represent 15 mi.

   d. _____ in. represent 30 mi.

   e. _____ in. represents $2\frac{1}{2}$ mi.

5. The area of Rhode Island is 1,545 square miles. The area of Delaware is 2,489 square miles. How much greater is the area of Delaware than of Rhode Island?

   _____ square miles

6. Solve the riddle.

   I am a polygon.

   I have two right angles.

   I have only one pair of parallel sides.

   What am I? _____

Date Time

Logic Problems

Math Message

1. There are three children in the Smith family: Sara, Sam, and Sue.
Use the following clues to find each one's age:

- Each of the two younger children is half as old as the next older child.
- The oldest is 16.
- Sara is not the oldest.
- Sara is twice as old as Sam.

What is the age of each person?

Sara ______ Sam ______ Sue ______

2. a. DeeAnn, Eric, Brooke, and Kelsey all have a favorite sport. Each one likes a different sport. Their favorite sports are basketball, swimming, golf, and tennis.

- DeeAnn doesn't like water.
- Both Eric and Brooke like to hit a ball.
- Eric doesn't like to play on a playing field that has lines on it.

What is each person's favorite sport?

DeeAnn ______ Eric ______

Brooke ______ Kelsey ______

b. Write an explanation of how your group found the answers.

______

______

______

______

______

______

Date Time

Logic Problems (cont.)

3. Raoul, Martha, Kwan, and Karen like to draw. One of them likes working with colored markers best, another with watercolor paints, another with colored chalk, and another with colored pencils.

- Raoul does not like to work with paintbrushes.
- Martha likes to sharpen her drawing tools.
- Kwan and Karen do not like dust.
- Karen sometimes gets bristles in her artwork.

Find out what each child likes best. Use the logic grid to help you.

	Colored markers	**Watercolor paints**	**Colored chalk**	**Colored pencils**
Raoul				
Martha				
Kwan				
Karen				

Raoul ______________ Martha ______________

Kwan ______________ Karen ______________

Date Time

Logic Problems (cont.)

4. Sam, Don, Darla, Jon, and Sara all have a favorite kind of cookie. They each like a different kind best.

- Sam and Jon do not like peanut butter.
- Don has never tried sugar cookies and neither has Sara.
- Darla does not like raisins.
- Jon doesn't like sugar cookies.
- Darla and Jon do not like chocolate.
- Sara does like chocolate.
- Don likes cinnamon.

What kind of cookie does each like best? Use the logic grid to help you.

	Peanut butter	**Sugar**	**Cinnamon**	**Oatmeal raisin**	**Chocolate**
Sam					
Don					
Darla					
Jon					
Sara					

Sam ______________ Don ______________

Darla ______________ Jon ______________

Sara ______________

Date Time

Estimating Distances

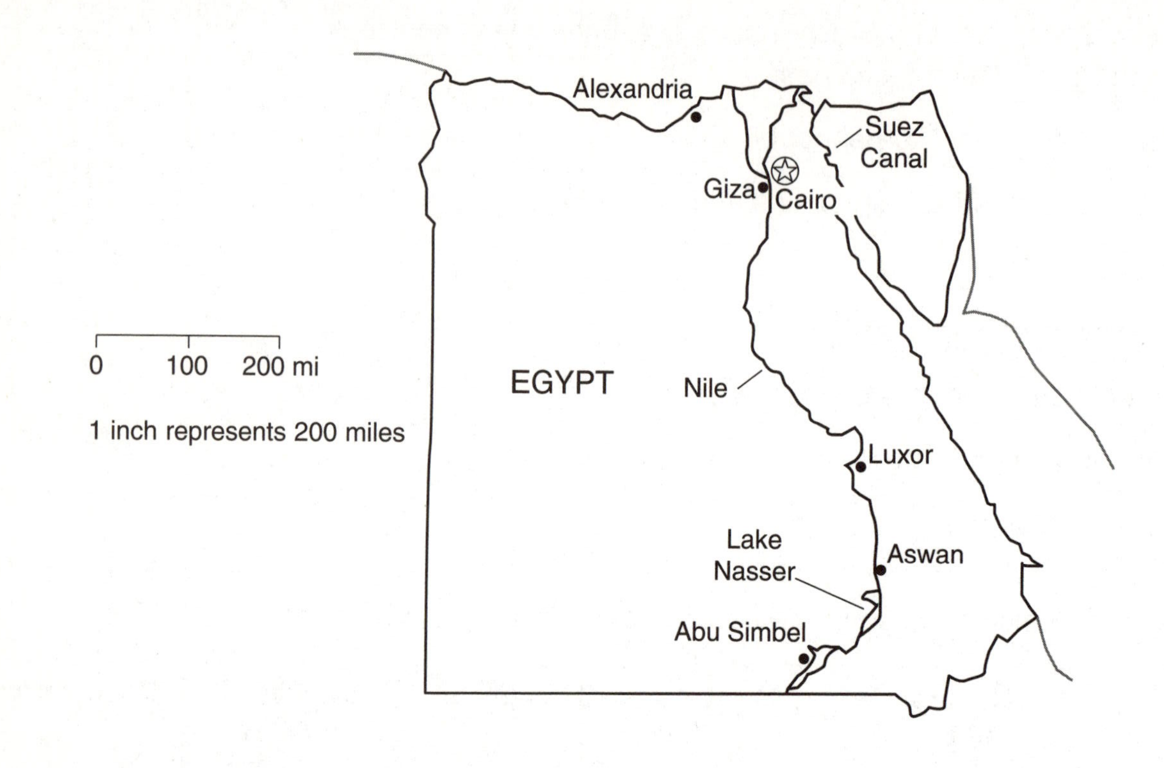

You want to take a trip to Egypt and see the following sights:

- Cairo, the capital, on the Nile River, near the Pyramids at Giza
- Alexandria, a busy modern city and port on the Mediterranean
- the Aswan High Dam across the Nile River, completed in 1970, and Lake Nasser, which formed behind the dam
- the temples at Abu Simbel, built more than 3,000 years ago, which were moved to their present location in the 1960s to escape the rising water of Lake Nasser

You want to know how far it is between locations.

1. The distance between Alexandria and Abu Simbel is about _______ inches on the map.

That is about _______ miles.

2. The distance between Cairo and Aswan is about _______ inches on the map.

That is about _______ miles.

3. The distance between Abu Simbel and Aswan is about _______ inch(es) on the map.

That is about _______ miles.

Date Time

Math Boxes 3.11

1. a. Measure the line segment to the nearest centimeter.

L ________________________________ P

_______ cm

b. Draw a line segment that is half the length of $\overline{LP}$.

c. How long is the line segment you drew? _______ cm

2. There were 123 recipients of the Congressional Medal of Honor for World War I and 433 for World War II. How many of these medals in all were issued for the two wars?

_______ medals

3. Complete the multiplication facts.

a. 4 * 7 = _______

b. 7 * 7 = _______

c. 5 * 7 = _______

d. _______ = 3 * 6

e. _______ = 6 * 5

f. _______ = 9 * 6

4. Tell whether each number sentence is true or false.

a. (7 * 4) – 2 = 80 – (9 * 6) _______

b. 8 * 11 = (8 * 5) + (8 * 6) _______

c. 34 < (3 * 5) + (63 / 9) _______

d. 12 = (6 * 7) / (19 – 7) _______

5. Make a true sentence by inserting parentheses.

a. 7 * 4 – 4 = 0

b. 45 / 9 + 10 = 15

c. 8 * 7 – 6 = 8

d. 24 / 3 + 5 = 13

Date Time

Time to Reflect

1. Why are multiplication turn-around facts called "turn-around facts"?

2. In this unit, your teacher encouraged you to use "A Guide for Solving Number Stories" when solving problems. Do you think the steps and suggestions in the guide are useful? Why or why not?

3. Is the World Tour Project a good way to learn about numbers? Is it helping you? Why or why not?

Use with Lesson 3.12.

Date Time

Math Boxes 3.12

1. Put these numbers in order from smallest to largest.

0.8 0.08 0.73 0.095

______

______

______

______

2. What numbers come next?

a. 0.1, 0.9, 1.7, ______, ______, ______

b. 5.6, 5.3, 5, ______, ______, ______

c. 8, 6.9, 5.8, ______, ______, ______

d. 7.23, 7.28, 7.33, ______, ______, ______

3. Add or subtract.

a. $11.03 + $4.79 = ______

b. $3.62 + $7.25 = ______

c. $8.96 − $5.92 = ______

d. $3.56 − $2.89 = ______

4. Blake estimated the length of his little finger. Circle the most reasonable estimate.

20 millimeters

45 millimeters

80 millimeters

5. a. Measure the line segment to the nearest centimeter.

A B

______ cm

b. Draw a line segment that is half the length of $\overline{AB}$.

c. How long is the line segment you drew? ______ cm

Date Time

Tenths and Hundredths

Base-10 Block	Symbol	Value
Flat	□	1
Long	|	$\frac{1}{10}$, or 0.1
Cube	▪	$\frac{1}{100}$, or 0.01

1. Complete the table.

Base-10 Blocks	Fraction Notation	Decimal Notation
||	$\frac{2}{10}$	0.2
▪▪▪▪▪▪▪▪▪		
||||| ||||		
□ ||| ▪▪▪▪▪▪		

2. Write each number in decimal notation.

Example $\frac{3}{10}$ = 0.3

a. $\frac{4}{10}$ = ______ **b.** $\frac{71}{100}$ = ______ **c.** $32\frac{6}{100}$ = ______

3. Draw base-10 blocks to show each number. Draw as few blocks as possible.

Example 0.3 |||

a. 0.43 **b.** 2.16

4. Write each of the following in decimal notation.

a. 9 tenths ______ **b.** 82 hundredths ______ **c.** 38 and 5 tenths ______

Problem-Solving Practice

1. Fifteen fourth graders were asked how many blocks they lived from school. The results are displayed in the following graph:

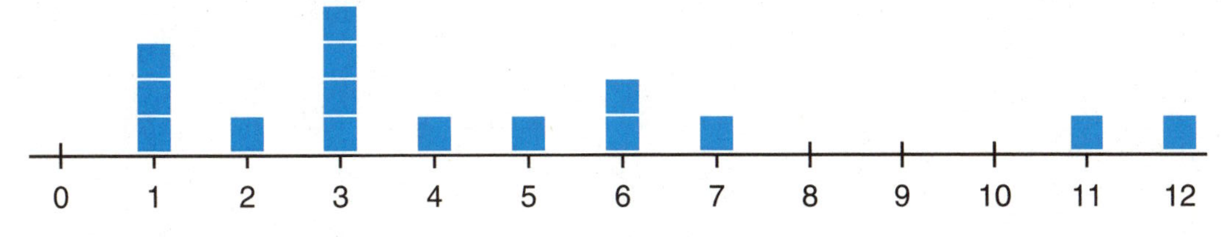

a. What is the maximum number of blocks any student lives from school? ________

b. What is the minimum number of blocks? ________

c. What is the range? ________

d. What is the mode? ________

e. What is the median? ________

For Problems 2–4, use the computation grid if you wish.

2. The Goldmans are taking a 350-mile trip. If they drive 185 miles before lunch, how much farther will they have to go?

Number model:

____

Answer: ________ miles

3. A doughnut costs 45 cents. Milk costs 35 cents a carton. How much will 2 doughnuts and 1 carton of milk cost?

Number model:

____

Answer: $________

4. 407 − 192 = ________

Date Time

Math Boxes 4.1

1. Which of the following is closest to the sum of 458, 1,999, and 12,307?

5,000

10,000

15,000

20,000

SRB 155 156

2. Complete the multiplication and division facts.

a. $6 * \_\_\_\_ = 18$

b. $3 * \_\_\_\_ = 21$

c. $16 \div 4 = \_\_\_\_$

d. $20 \div 5 = \_\_\_\_$

e. $54 \div 6 = \_\_\_\_$

f. $9 * 4 = \_\_\_\_$

SRB 258

3. Solve.

a. $323 - y = 42$ $y = \_\_\_\_$

b. $199 = p - 408$ $p = \_\_\_\_$

c. $r + 115 = 655$ $r = \_\_\_\_$

d. $430 + s = 700$ $s = \_\_\_\_$

SRB 128

4. In the numeral 59.378, the 5 stands for 5 tens.

a. The 9 stands for ________.

b. The 3 stands for ________.

c. The 7 stands for ________.

d. The 8 stands for ________.

SRB 28 29

5. Draw and label ray *BY*.

Draw point *A* on it.

SRB 77

6. Make true sentences by inserting parentheses.

a. $5 * 4 - 2 = 18$

b. $25 + 8 * 7 = 81$

c. $1 = 36 / 6 - 5$

d. $24 = 81 / 9 + 15$

SRB 129

Date Time

Comparing Decimals

Math Message

1. Arjun thought that 0.3 was less than 0.15. How could you help Arjun see that 0.3 is more than 0.15?

____

____

____

____

2. Use base-10 blocks to complete the following table.

"<" means "is less than."

">" means "is greater than."

Base-10 Blocks	Decimal	>, <, or =	Decimal	Base-10 Blocks
	0.2	>	0.12	
			0.1	
	0.13			
			0.3	
	1.2			
			0.39	
	2.3			

Date Time

Ordering Decimals

1. Write < or >.

a. 0.24 ________ 0.18 **b.** 0.05 ________ 0.1 **c.** 0.2 ________ 0.35

d. 1.03 ________ 0.30 **e.** 3.2 ________ 6.59 **f.** 25.9 ________ 25.72

2. Put these numbers in order from smallest to largest.

a. 0.05, 0.5, 0.55, 5.5 ________ ________ ________ ________
smallest largest

b. 0.99, 0.27, 1.8, 2.01 ________ ________ ________ ________
smallest largest

c. 2.1, 2.01, 20.1, 20.01 ________ ________ ________ ________
smallest largest

d. 0.01, 0.10, 0.11, 0.09 ________ ________ ________ ________
smallest largest

3. "What's green inside, white outside, and hops?"
To find the answer, put the numbers in order from smallest to largest.

0.66	1	0.2	1.05	0.90	0.01	0.75	0.35	$\frac{1}{4}$	$\frac{1}{2}$	0.05	0.09	5.5
N	I	O	C	W	A	D	S	G	A	F	R	H

Write your answers in the following table. The first answer is done for you.

0.01												
A												

Date Time

Math Boxes 4.2

1. Complete.

a. $5 * 3 =$ ________

b. $5 * 30 =$ ________

c. ________ $= 5 * 9$

d. ________ $= 50 * 9$

e. $8 * 9 =$ ________

f. $8 * 90 =$ ________

SRB 16

2. If $\frac{1}{4}$ inch on a map represents 40 miles, then

a. $\frac{1}{2}$ in. represents ________ mi.

b. 1 in. represents ________ mi.

c. $1\frac{3}{4}$ in. represent ________ mi.

d. 2 in. represent ________ mi.

e. $2\frac{1}{4}$ in. represent ________ mi.

SRB 125

3. Insert >, <, or =.

a. 0.96 ________ 0.4

b. 0.50 ________ 0.500

c. 1.3 ________ 1.09

d. 0.85 ________ 0.86

SRB 30 31

4. Add or subtract.

a. $\begin{array}{r} 391 \\ +\ 467 \\ \hline \end{array}$

b. $\begin{array}{r} 983 \\ -\ 494 \\ \hline \end{array}$

SRB 9-14

5. Ray, Samantha, Josh, and Erica each have a different favorite after-school snack: ice cream, pretzels, potato chips, or cookies. Find out which snack is the favorite for each child. (*Hint:* Use a logic grid if you need help.)

- Josh does not like salty snacks.
- Ray likes a snack that is cold and comes in many different flavors.
- Samantha likes a snack that is salty and may sometimes be twisted.

Ray ________ Samantha ________

Josh ________ Erica ________

Date Time

Uses of Decimals

Math Message

Describe two examples in which decimals are used in real life.

A Bicycle Trip

Bill and Alex often took all-day bicycle trips together.

During the summer, they took a 3-day bicycle tour. They carried camping gear in their saddlebags for the two nights they would be away from home.

Alex had a **trip meter** that showed miles traveled, in tenths of miles. He kept a log of the distances they traveled each day, before and after lunch.

Travel Log		
	Distance Traveled	
Timetable	Before lunch	After lunch
Day 1	27.0 mi	31.3 mi
Day 2	36.6 mi	20.9 mi
Day 3	25.8 mi	27.0 mi

Use estimation to answer the following questions. Do not work the problems on paper or with a calculator.

1. On which day did they travel the most miles? ____________

2. On which day did they travel the fewest miles? ____________

3. During the whole trip, did they travel more miles before or after lunch? ____________

4. Estimate the total distance they traveled. Circle your estimate.

   a. less than 150 miles
   b. between 150 and 180 miles
   c. between 180 and 200 miles
   d. more than 200 miles

5. On Day 1, about how many more miles did they travel after lunch than before lunch? ____________

6. Bill said that they traveled 1.2 more miles before lunch on Day 1 than on Day 3. Alex disagreed. He said they traveled 2.2 more miles. Who is right? ____________

Date Time

Math Boxes 4.3

1. Which of the following is closest to the sum of 3,005, 865, and 2,109?

5,000

10,000

15,000

20,000

2. Complete the multiplication and division facts.

a. 9 * ________ = 27

b. 5 * ________ = 40

c. 42 ÷ 6 = ________

d. 54 ÷ 9 = ________

e. 63 ÷ 7 = ________

f. 9 * 9 = ________

3. Solve.

a. $503 + y = 642$ $y =$ ________

b. $p + 263 = 319$ $p =$ ________

c. $r - 320 = 600$ $r =$ ________

d. $444 - s = 93$ $s =$ ________

4. In 34.561,

a. the 3 stands for ________________.

b. the 4 stands for ________________.

c. the 5 stands for ________________.

d. the 6 stands for ________________.

e. the 1 stands for ________________.

5. Draw and label ray *CT*.

Draw point *A* on it.

6. Make true sentences by inserting parentheses.

a. 7 * 8 – 6 = 50

b. 13 – 4 * 6 = 54

c. 10 = 49 / 7 + 3

d. 28 = 42 / 7 + 22

Date Time

Decimal Addition and Subtraction

Math Message

What's wrong with this problem? ____________________

What is the correct answer? ____________________

$$\begin{array}{r} 0.76 \\ +\ 0.2 \\ \hline 0.78 \end{array}$$

____

____

Decimal Addition and Subtraction

Add or subtract. Show your work below.

1. 2.05 + 1.83 = ________
2. 3.04 – 2.8 = ________
3. 1 – 0.67 = ________
4. 2.4 + 3.01 + 0.26 = ________
5. 2.31 – 1.88 = ________
6. 19 + 1.9 = ________

Date Time

Decimals on Grids

This grid is the whole.

Fractions and decimals can be shown by shading the grid.

Example

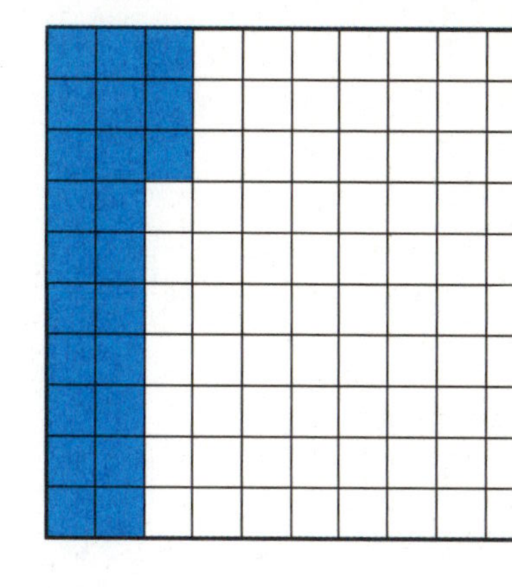

0.23, or $\frac{23}{100}$

Write a decimal and a fraction for each grid.

1.

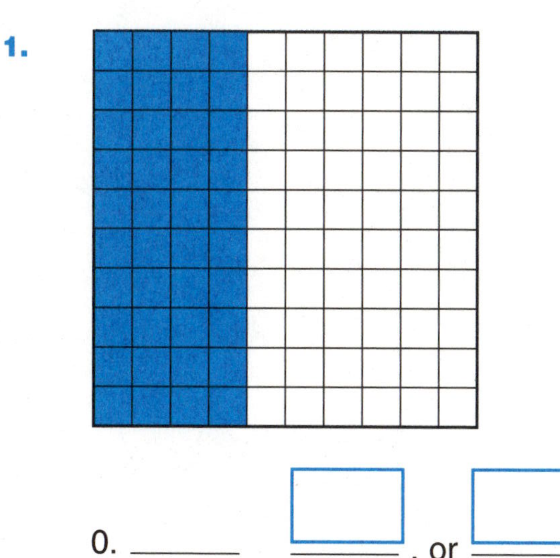

0. ______ $\frac{\square}{\square}$, or $\frac{\square}{\square}$

2.

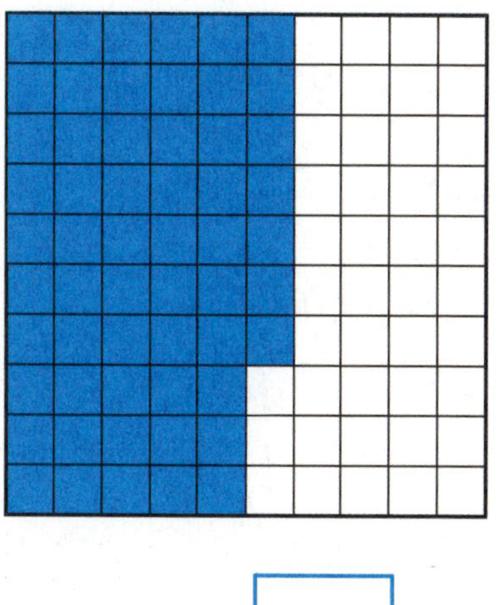

0. ______

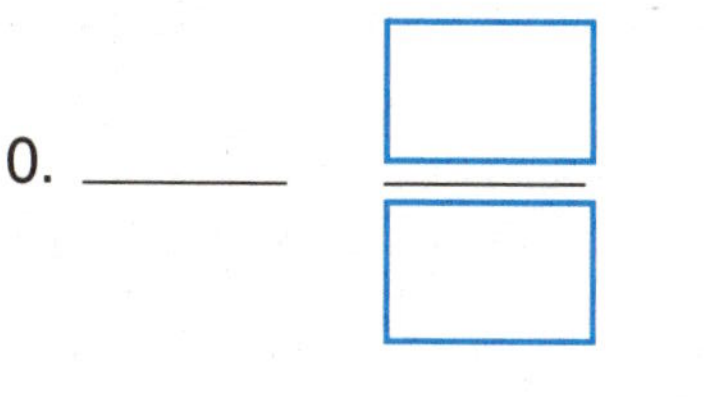

Shade each grid to show the decimal.

3. 0.32

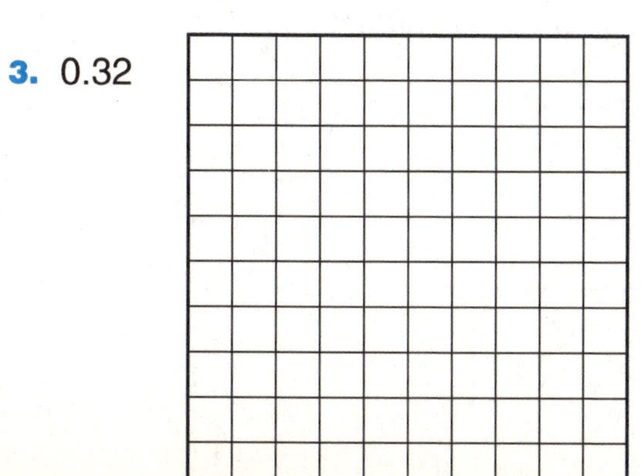

4. 0.05

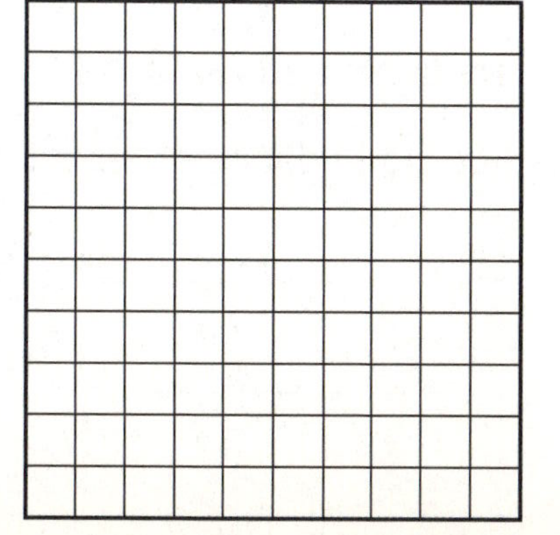

Date Time

Math Boxes 4.4

1. Complete.

a. $6 * 9 =$ ________

b. $6 * 90 =$ ________

c. ________ $= 5 * 8$

d. ________ $= 50 * 8$

e. $4 * 4 =$ ________

f. $4 * 40 =$ ________

2. If 2 centimeters on a map represent 50 kilometers, then

a. 1 cm represents ________ km.

b. 3 cm represent ________ km.

c. 4 cm represent ________ km.

d. 7 cm represent ________ km.

e. 8 cm represent ________ km.

3. Insert >, <, or =.

a. 0.6 ________ 0.57

b. 0.37 ________ 0.36

c. 2.56 ________ 2.056

d. 0.24 ________ 0.240

4. Add or subtract.

a. $\begin{array}{r} 814 \\ +\ 123 \\ \hline \end{array}$

b. $\begin{array}{r} 754 \\ -\ 396 \\ \hline \end{array}$

5. Bill, Marta, Monique, and Louis each participate in a different after-school activity: hockey, dance, gymnastics, or swimming. Find out in which activity each child participates.

- Louis wears a lot of equipment for protection during his after-school activity.
- Marta wears tights and leotards for her after-school activity.
- Bill does not like to dance or be in the water.

Bill ________ Marta ________

Monique ________ Louis ________

Date Time

Using Decimals

Math Message

Solve. Show your work below.

1. $18.09 + $7.24 = $__________

2. $3.78 + $5.22 = $__________

3. $24.61 − $19.38 = $__________

4. $12 − $1.47 = $__________

Keeping a Bank Balance

On January 2, Kate's aunt opened a bank account for Kate. Her aunt deposited $100.00 in the account.

Over the next several months, Kate made regular deposits into her account. She deposited part of her allowance and most of the money she made babysitting.

Kate also made a few withdrawals—to buy a radio and some clothes.

Think about the answers to the following questions:

- When you **withdraw** money, do you take money out or put money in?
- When you **deposit** money, do you take money out or put money in?
- When your money earns **interest,** does this add money to your account or take money away?

Date Time

Keeping a Bank Balance (cont.)

The following table shows the transactions (deposits and withdrawals) that Kate made during the first 4 months of the year and the interest earned.

1. In March, Kate took more money out of her bank account than she put in. In which other month did she withdraw more money than she deposited? ______

2. Complete the table. Remember to **add** if Kate makes a deposit or earns interest and to **subtract** if she makes a withdrawal.

Date	Transaction		Current Balance
January 2	Deposit	$100.00	$ 100.00
January 14	Deposit	$14.23	+ $ 14.23 $ 114.23
February 4	Withdrawal	$16.50	$ ______ $ ______
February 11	Deposit	$33.75	$ ______ $ ______
February 14	Withdrawal	$16.50	$ ______ $ ______
March 19	Deposit	$62.00	$ ______ $ ______
March 30	Withdrawal	$104.26	$ ______ $ ______
March 31	Bank credited Kate's account with $0.78 in interest		$ ______ $ ______
April 1	Deposit	$70.60	$ ______ $ ______
April 3	Withdrawal	$45.52	$ ______ $ ______
April 28	Withdrawal	$27.91	$ ______ $ ______

Check your answers with a partner.

Forming a Relay Team

Mrs. Wong, the gym teacher, wants to form 3 teams for a 200-yard relay race. There will be 4 students on each team. Each student will run 50 yards.

The table at the right shows how long it took some fourth-grade students to run 50 yards the last time they had a race. They were timed to the nearest tenth of a second.

Mrs. Wong will use these times to try to predict about how long it would take various combinations of students to run the 200-yard relay race.

Here are some combinations she tried. If she uses the times shown in the table, about how long (to the nearest tenth of a second) would the following teams take to run the 200-yard relay race?

Runner	Time (seconds)
Art	6.3
Bruce	7.0
Jamal	7.4
Doug	7.9
Al	8.3
Will	8.8
Linda	6.2
Sue	7.6
Pat	7.7
Mary	8.1
Alba	8.4
Joyce	8.5

1. Art, Bruce, Jamal, Doug — About ____.____ seconds
2. the 4 fastest girls — About ____.____ seconds
3. the 4 slowest students — About ____.____ seconds
4. the 4 fastest students — About ____.____ seconds
5. the 2 fastest and the 2 slowest students — About ____.____ seconds
6. About how many yards per second did Bruce run? ____ yards per second

Challenge

7. Make up 3 teams that will be fairly evenly matched.

Estimated time

Team 1: ______________________ About ____.____ seconds

Team 2: ______________________ About ____.____ seconds

Team 3: ______________________ About ____.____ seconds

Date ________ Time ________

Math Boxes 4.5

1. Add.

a.
$$\begin{array}{r} 6 \\ 40 \\ 150 \\ +\ 1{,}000 \\ \hline \end{array}$$

b.
$$\begin{array}{r} 54 \\ 180 \\ 240 \\ +\ 800 \\ \hline \end{array}$$

SRB 9 10

2. Complete the multiplication and division facts.

a. 7 * ________ = 28

b. 48 ÷ ________ = 8

c. ________ ÷ 8 = 7

d. ________ * 10 = 50

e. 8 * ________ = 64

3. Solve each open sentence.

a. $9.4 - K = 3$ $K =$ ________

b. $2.34 = S + 1.06$ $S =$ ________

c. $R - 12.2 = 4.65$ $R =$ ________

d. $0.81 - M = 0.43$ $M =$ ________

e. $F - 2.1 = 6.8$ $F =$ ________

SRB 128

4. Add 9 tens, 8 hundredths, and 3 tenths to 34.53.

What is the result? ________

SRB 34

5. Write as dollars and cents.

a. 20 dimes = $______.______

b. 20 nickels = $______.______

c. 20 quarters = $______.______

d. 10 quarters and 7 dimes =

$______.______

6. Make up a set of 5 numbers having the following landmarks:

median: 8

range: 10

minimum: 3

______, ______, ______, ______, ______

SRB 65

Date Time

Tenths, Hundredths, and Thousandths

Math Message

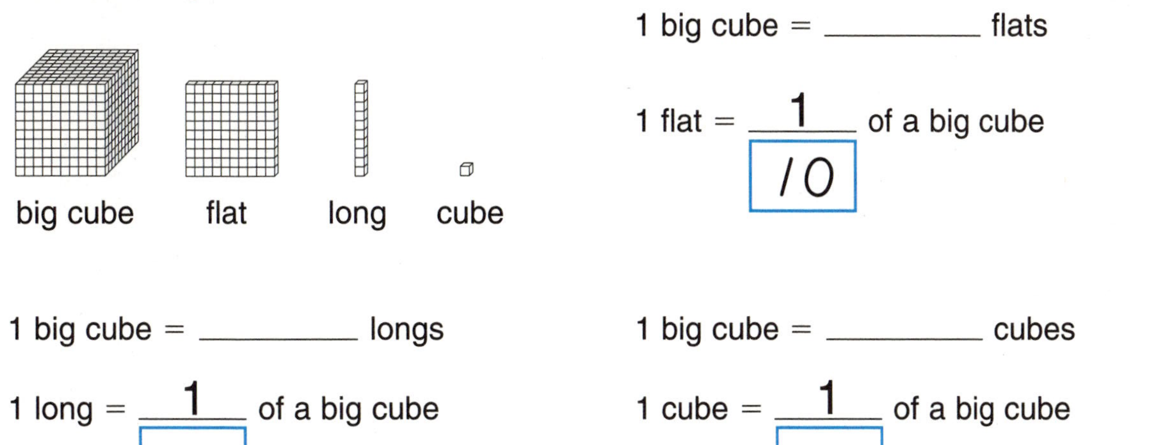

1 big cube = ________ flats

1 flat = $\frac{1}{\boxed{10}}$ of a big cube

1 big cube = ________ longs

1 long = $\frac{1}{\boxed{}}$ of a big cube

1 big cube = ________ cubes

1 cube = $\frac{1}{\boxed{}}$ of a big cube

Base-10 Block	Symbol	Value
Big Cube	⧉	1
Flat	□	$\frac{1}{10}$, or 0.1
Long	|	$\frac{1}{100}$, or 0.01
Cube	▪	$\frac{1}{1,000}$, or 0.001

1. Complete the table.

Base-10 Blocks	Fraction Notation	Decimal Notation
□ |||| ▪▪	$\frac{142}{1,000}$	0.142
□□ ▪▪▪▪▪		
||||| |		
□□□□□		
⧉ |||| ▪▪▪		

Tenths, Hundredths, and Thousandths (cont.)

2. Write each number in decimal notation.

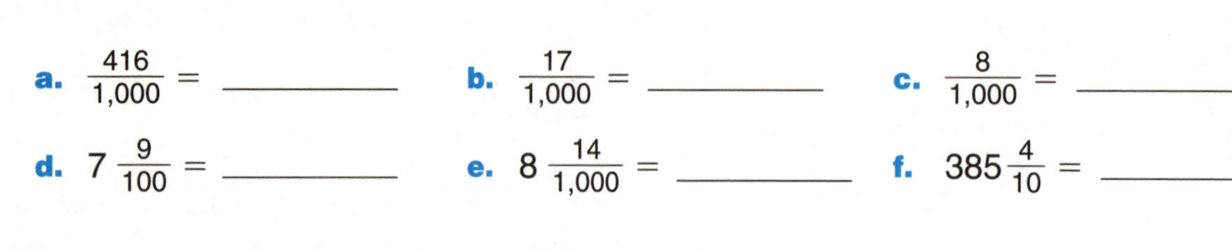

Example $\frac{72}{1,000}$ = 0.072

a. $\frac{416}{1,000}$ = ______ **b.** $\frac{17}{1,000}$ = ______ **c.** $\frac{8}{1,000}$ = ______

d. $7\frac{9}{100}$ = ______ **e.** $8\frac{14}{1,000}$ = ______ **f.** $385\frac{4}{10}$ = ______

3. Draw base-10 blocks to show each number.

Example 0.205

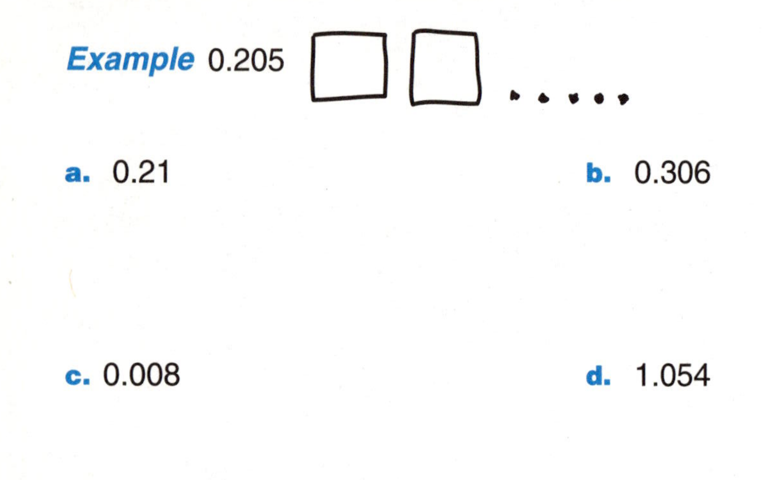

a. 0.21 **b.** 0.306

c. 0.008 **d.** 1.054

4. Write each of the following in decimal notation.

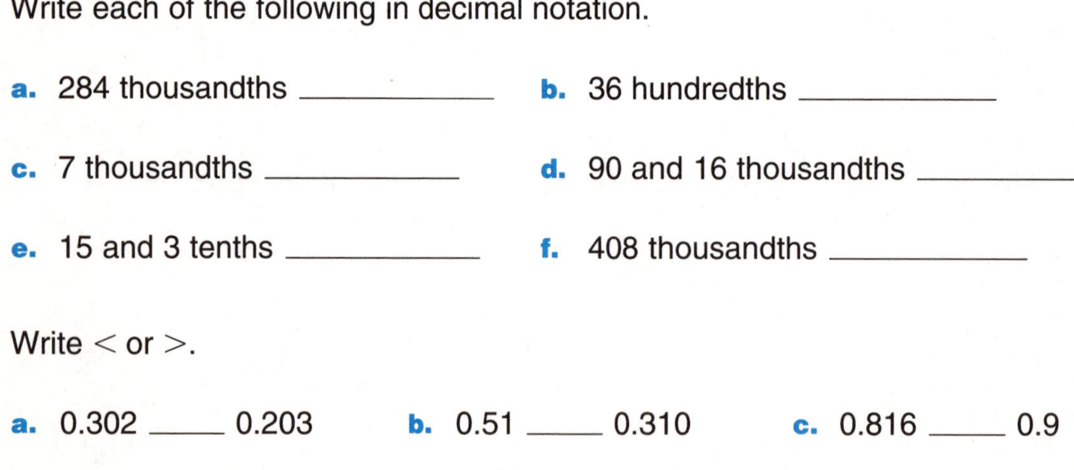

a. 284 thousandths ______ **b.** 36 hundredths ______

c. 7 thousandths ______ **d.** 90 and 16 thousandths ______

e. 15 and 3 tenths ______ **f.** 408 thousandths ______

5. Write $<$ or $>$.

a. 0.302 ____ 0.203 **b.** 0.51 ____ 0.310 **c.** 0.816 ____ 0.9

d. 1.47 ____ 1.5 **e.** 0.073 ____ 0.73 **f.** 4.01 ____ 4.009

Date Time

Math Boxes 4.6

1. Write the following numbers using digits:

a. three million, four hundred thirty-two thousand, nine ____________

b. six-hundred million, five thousand, twenty-one ____________

SRB 4

2. a. Measure the length of this line segment to the nearest centimeter.

About ______ cm

b. Draw a line segment 3 centimeters long.

SRB 108

3. Put these numbers in order from smallest to largest.

5.92 0.95 9.25 2.95 0.92

_____ _____ _____ _____ _____

SRB 30 31

4. A trumpeter swan can weigh about 16.8 kilograms. A Manchurian crane can weigh about 14.9 kilograms. How much heavier is a trumpeter swan than a Manchurian crane?

______ kilograms

SRB 132 152

5. Add or subtract.

a. $5.18 - 3.65 =$ ______

b. $11.2 - 3.9 =$ ______

c. $16.86 + 9.24 =$ ______

d. $0.87 + 0.94 =$ ______

SRB 32-35

Date Time 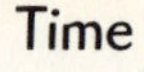

Measuring Length with Metric Units

Your teacher will assign you several objects or distances to measure. Measure each to the nearest centimeter. Then compare your measurements with your partner's. If you do not agree, work together to measure the object again. Record the results in the table.

Object or Distance	My Measurement	Partner's Measurement	Agreed Measurement
	About ______ cm	About ______ cm	About ______ cm
	About ______ cm	About ______ cm	About ______ cm
	About ______ cm	About ______ cm	About ______ cm
	About ______ cm	About ______ cm	About ______ cm
	About ______ cm	About ______ cm	About ______ cm

Number Lines

Fill in the missing numbers.

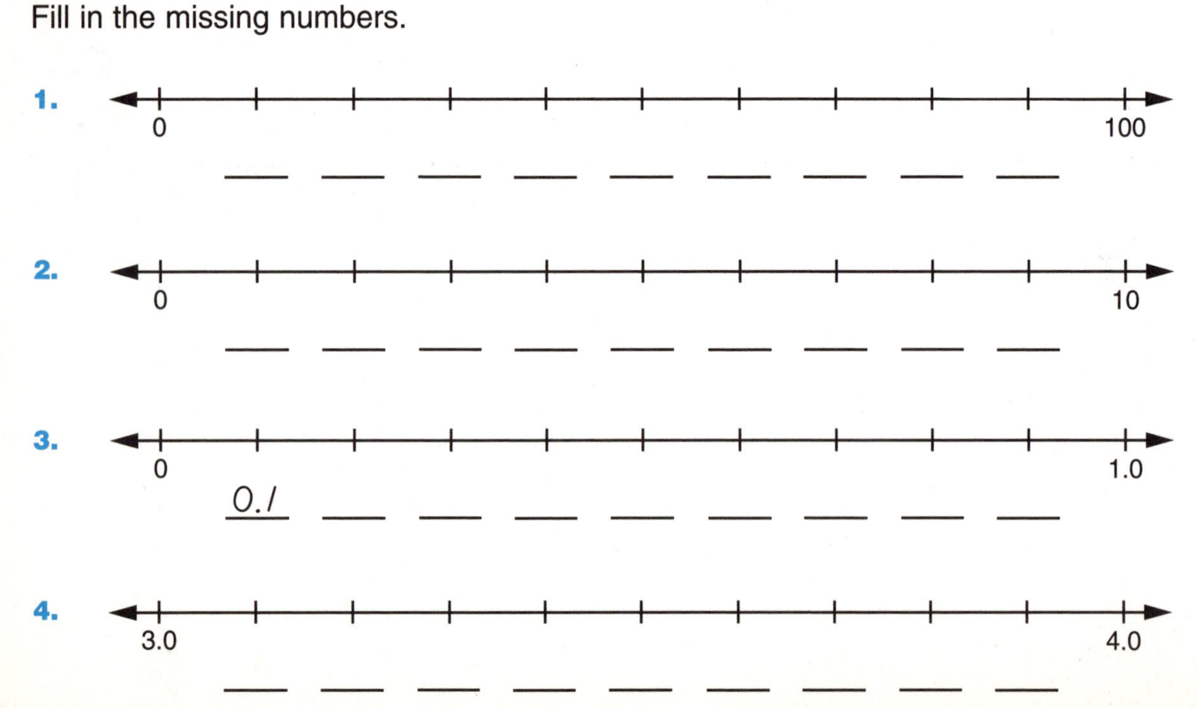

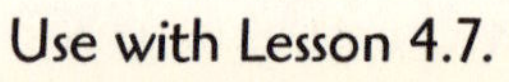

Date Time

Math Boxes 4.7

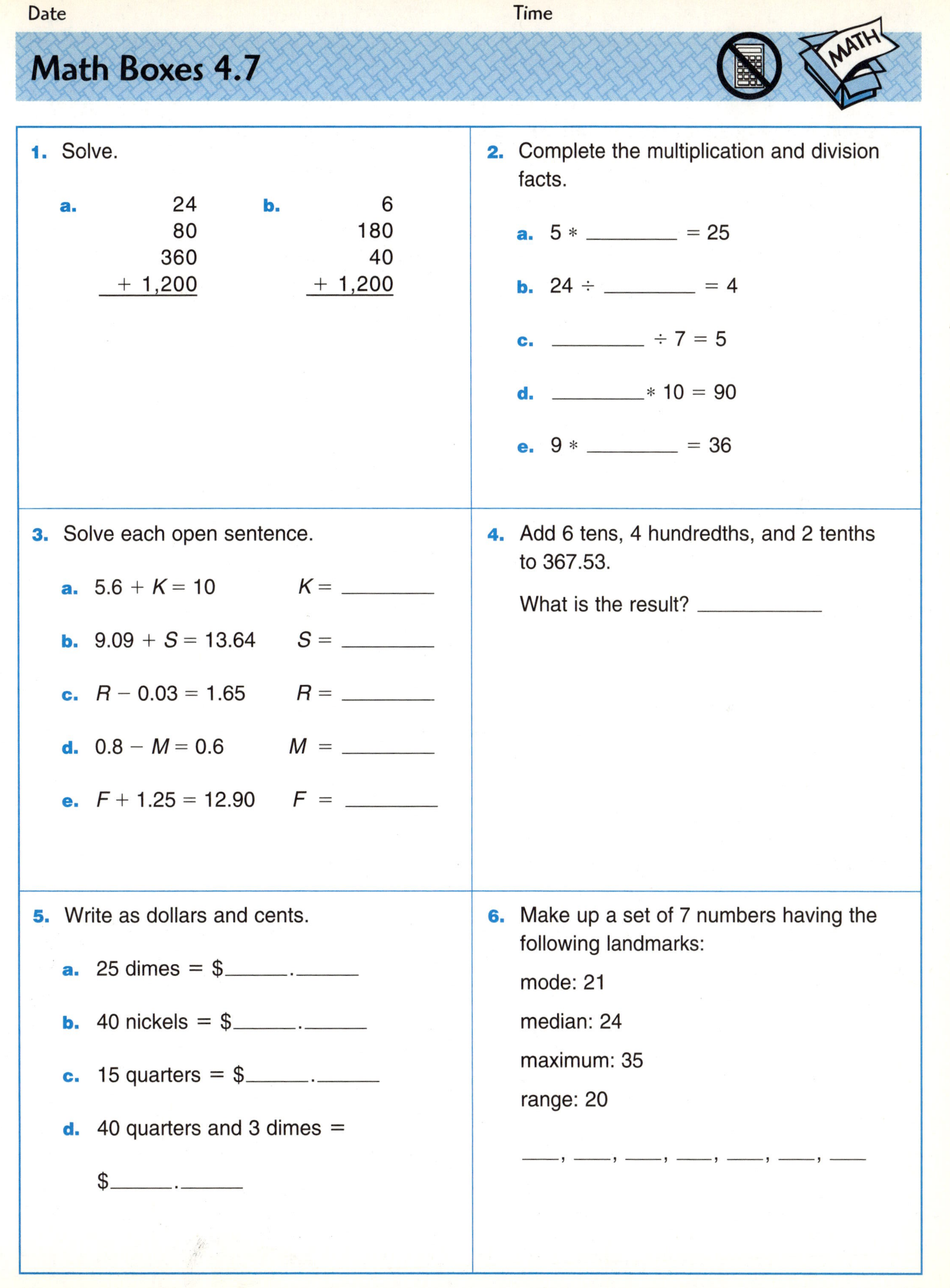

1. Solve.

a.
$$\begin{array}{r} 24 \\ 80 \\ 360 \\ +\ 1{,}200 \\ \hline \end{array}$$

b.
$$\begin{array}{r} 6 \\ 180 \\ 40 \\ +\ 1{,}200 \\ \hline \end{array}$$

2. Complete the multiplication and division facts.

a. 5 * ________ = 25

b. 24 ÷ ________ = 4

c. ________ ÷ 7 = 5

d. ________ * 10 = 90

e. 9 * ________ = 36

3. Solve each open sentence.

a. $5.6 + K = 10$ K = ________

b. $9.09 + S = 13.64$ S = ________

c. $R - 0.03 = 1.65$ R = ________

d. $0.8 - M = 0.6$ M = ________

e. $F + 1.25 = 12.90$ F = ________

4. Add 6 tens, 4 hundredths, and 2 tenths to 367.53.

What is the result? ____________

5. Write as dollars and cents.

a. 25 dimes = $______.______

b. 40 nickels = $______.______

c. 15 quarters = $______.______

d. 40 quarters and 3 dimes =

$______.______

6. Make up a set of 7 numbers having the following landmarks:

mode: 21

median: 24

maximum: 35

range: 20

___, ___, ___, ___, ___, ___, ___

Date Time

Personal References for Units of Length

Personal References for Metric Units of Length

Use a ruler, meterstick, or tape measure to find common objects that have lengths of 1 centimeter, 1 decimeter, and 1 meter. The lengths do not have to be exact, but they should be close. Ask a friend to look for references with you. You can find more than one reference for each unit. Record the references in the table below.

Unit of Measure	Personal References
1 centimeter (cm)	
1 decimeter (dm), or 10 centimeters	
1 meter (m)	

To be completed in Lesson 5.1.
Personal References for U.S. Customary Units of Length

Use a ruler, yardstick, or tape measure to find common objects that have lengths of 1 inch, 1 foot, and 1 yard. The lengths do not have to be exact, but they should be close. Ask a friend to look for references with you. You can find more than one reference for each unit. Record the references in the table below.

Unit of Measure	Personal References
1 inch (in.)	
1 foot (ft)	
1 yard (yd)	

Date Time

My Measurement Collection for Metric Units of Length

Use your personal references to estimate the length of an object or a distance in centimeters, decimeters, or meters. Describe the object or distance and record your estimate in the table below. Then measure the object or distance and record the actual measurement in the table.

Object or Distance	Estimated Length	Actual Length

Date ____________ Time ____________

Math Boxes 4.8

1. a. Is 47 closer to 40 or 50?

____

b. Name the number halfway between 30 and 40. ________

SRB

2. Without measuring, estimate the length of your foot, from heel to toe. Then measure the length of your foot.

a. Estimate: About ________ cm

b. Measurement: About ________ cm

SRB 108 110

3. Complete.

a. 1 cm = ________ mm

b. 5 cm = ________ mm

c. ________ cm = 30 mm

d. 100 cm = ________ m

e. 200 cm = ________ m

f. 1 m = ________ mm

SRB 109

4. Add or subtract.

a. $\begin{array}{r} 309 \\ +\ 721 \\ \hline \end{array}$

b. $\begin{array}{r} 700 \\ -\ 299 \\ \hline \end{array}$

SRB 9-14

5. Tell whether each number sentence is true or false.

a. $8.77 - 0.08 = 8.50$ ________

b. $35.7 + 22.1 = 57.87$ ________

c. $90.2 - 44.9 < 45$ ________

d. $4.66 + 2.13 > 6$ ________

SRB 34 35 128

6. Solve the riddle.

I am a polygon.

All my angles have equal measures.

All my sides have equal measures.

What am I? ____________________

SRB 83

Date Time

Measuring in Millimeters

Math Message

On your centimeter ruler, the numbered marks are for centimeters and the little marks between centimeter marks are for millimeters.

1. Look at your centimeter ruler. How many millimeters are in 1 centimeter? _______ mm

2. Name something that measures about 1 millimeter. ____________________________

____

____

3. Draw a line segment that is 8 centimeters long.

4. Draw a second line segment that is 80 millimeters long.

Measure each line segment below, using both the millimeter side and the centimeter side of the cm/mm ruler. Record both measurements.

5. A ______________________________ B

   Length of $\overline{AB}$ = _______ mm = _______ cm

6. C _______________ D Length of $\overline{CD}$ = _______ mm = _______ cm

7. E __ F Length of $\overline{EF}$ = _______ mm = _______ cm

---

Measuring Land Invertebrates

An **invertebrate** is an animal that does not have a backbone. (The backbone is also called the spinal column.) Some invertebrates live on land, others in the water. The most common land invertebrates are insects.

The invertebrates shown on page 104, except the earthworm, bumblebee, and mealybug, have been drawn to about actual size. The earthworm can grow to about 4 times the length shown. The bumblebee is shown about twice its actual size and the mealybug about 3 times its actual size.

Date Time

Measuring Land Invertebrates (cont.)

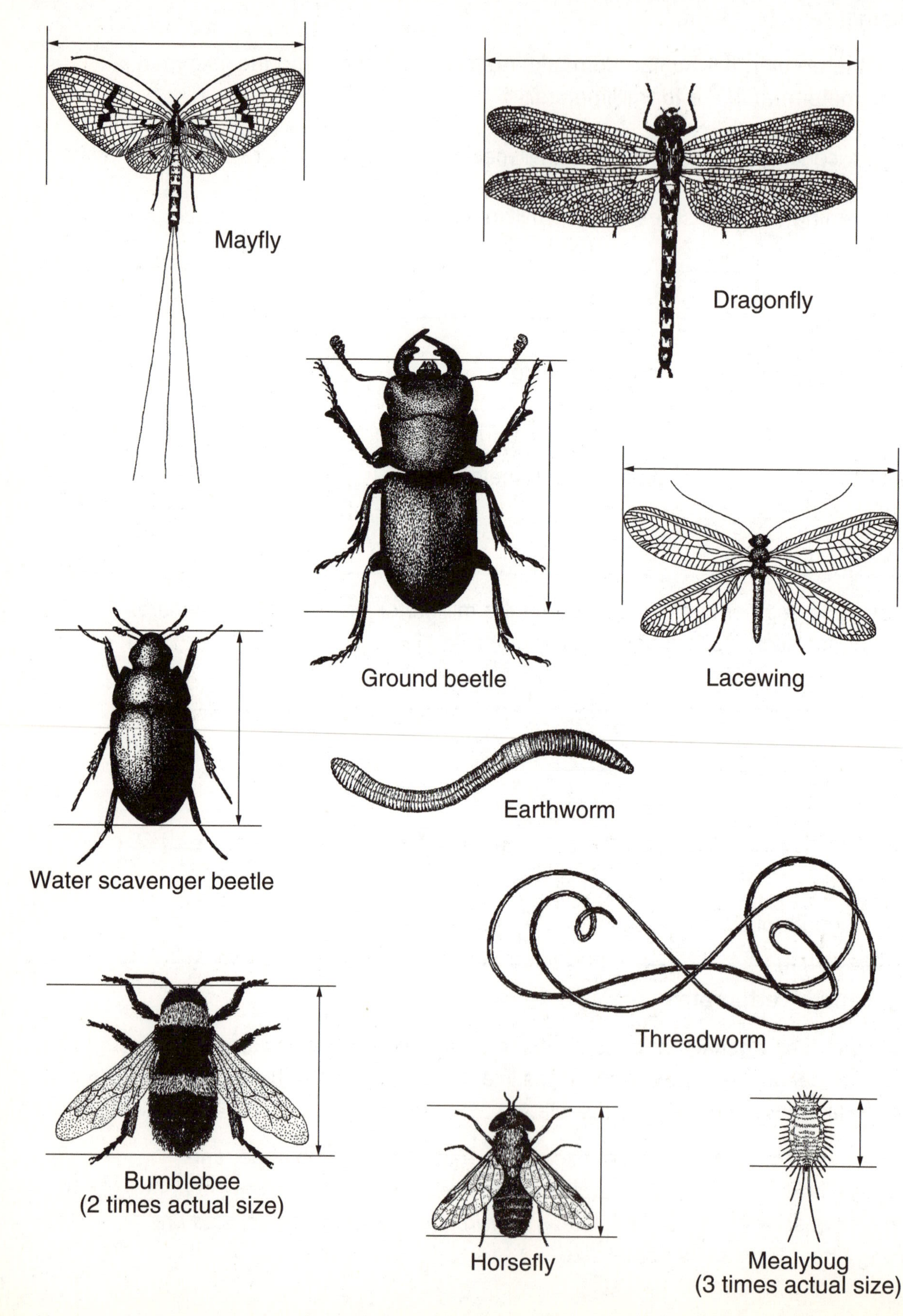

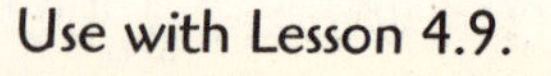

Date Time

Measuring Land Invertebrates (cont.)

Refer to the pictures on page 104 to answer the following questions.

1 centimeter (cm) = 10 millimeters (mm)
1 millimeter = 0.1 centimeter

1. Measure the following invertebrates to the nearest millimeter by finding the distance between the two guidelines. Then give the lengths in centimeters.

a. mayfly	About ______ mm	About ______ cm
b. dragonfly	About ______ mm	About ______ cm
c. water scavenger beetle	About ______ mm	About ______ cm
d. ground beetle	About ______ mm	About ______ cm
e. lacewing	About ______ mm	About ______ cm
f. horsefly	About ______ mm	About ______ cm

2. How much longer is the ground beetle than the water scavenger beetle? About ______ cm

3. The bee has been drawn to twice its actual size. In reality, which is longer, the bee or the horsefly? ______________ How much longer? About ______ mm

4. The mealybug has been drawn to 3 times its actual size. In the space at the right, draw a mealybug that is about the actual size.

5. What is the actual size of the mealybug in millimeters? ______ mm

6. When straight, the threadworm in the drawing is 306 millimeters long.

   What is its length in centimeters? ______ cm In meters? ______ m

Date ____ Time ____

Math Boxes 4.9

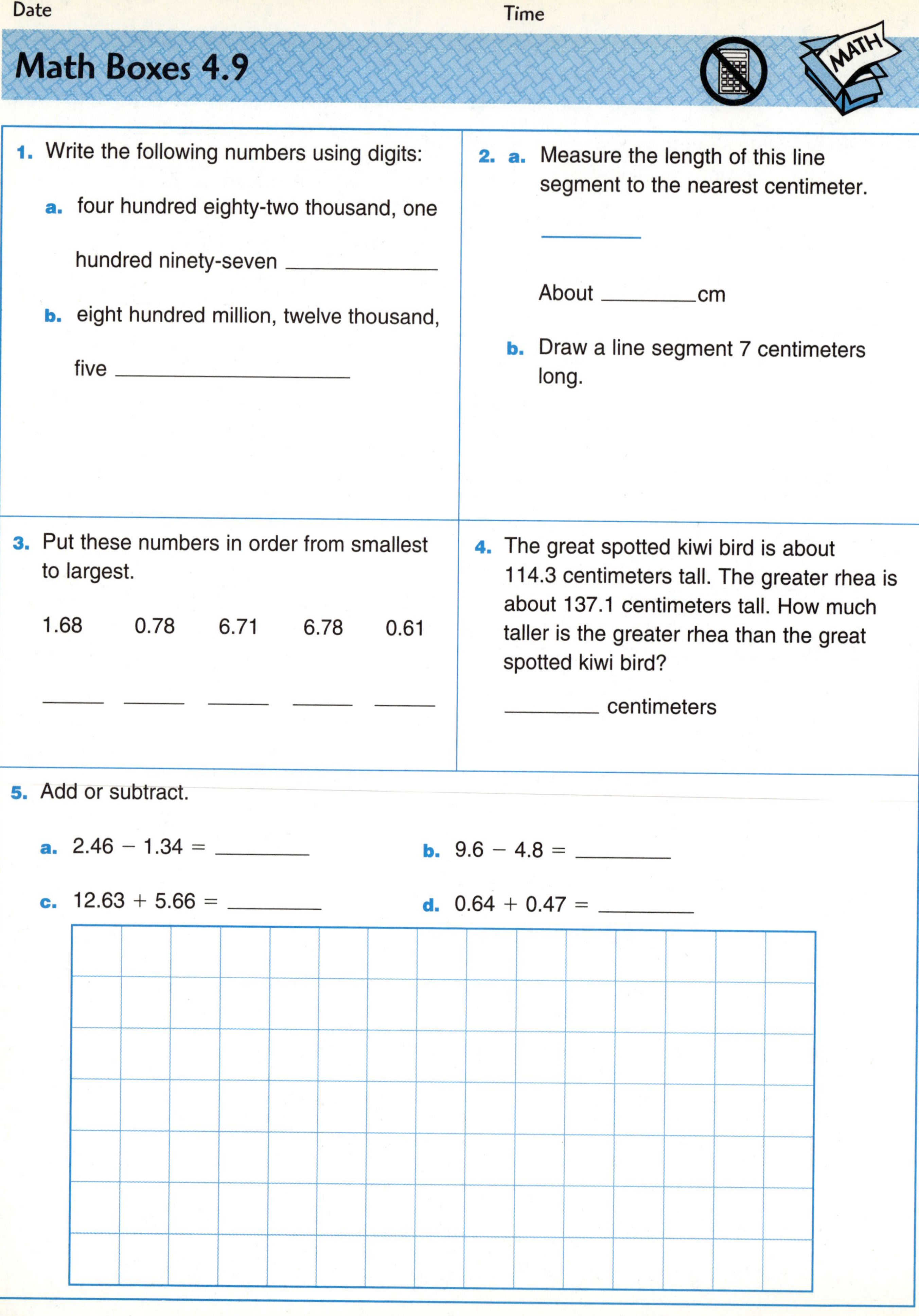

1. Write the following numbers using digits:

   a. four hundred eighty-two thousand, one hundred ninety-seven ______

   b. eight hundred million, twelve thousand, five ______

2. a. Measure the length of this line segment to the nearest centimeter.

   About ______ cm

   b. Draw a line segment 7 centimeters long.

3. Put these numbers in order from smallest to largest.

   1.68   0.78   6.71   6.78   0.61

   ____ ____ ____ ____ ____

4. The great spotted kiwi bird is about 114.3 centimeters tall. The greater rhea is about 137.1 centimeters tall. How much taller is the greater rhea than the great spotted kiwi bird?

   ______ centimeters

5. Add or subtract.

   a. $2.46 - 1.34 =$ ______

   b. $9.6 - 4.8 =$ ______

   c. $12.63 + 5.66 =$ ______

   d. $0.64 + 0.47 =$ ______

Date Time

Place-Value Number Lines

Fill in the missing numbers.

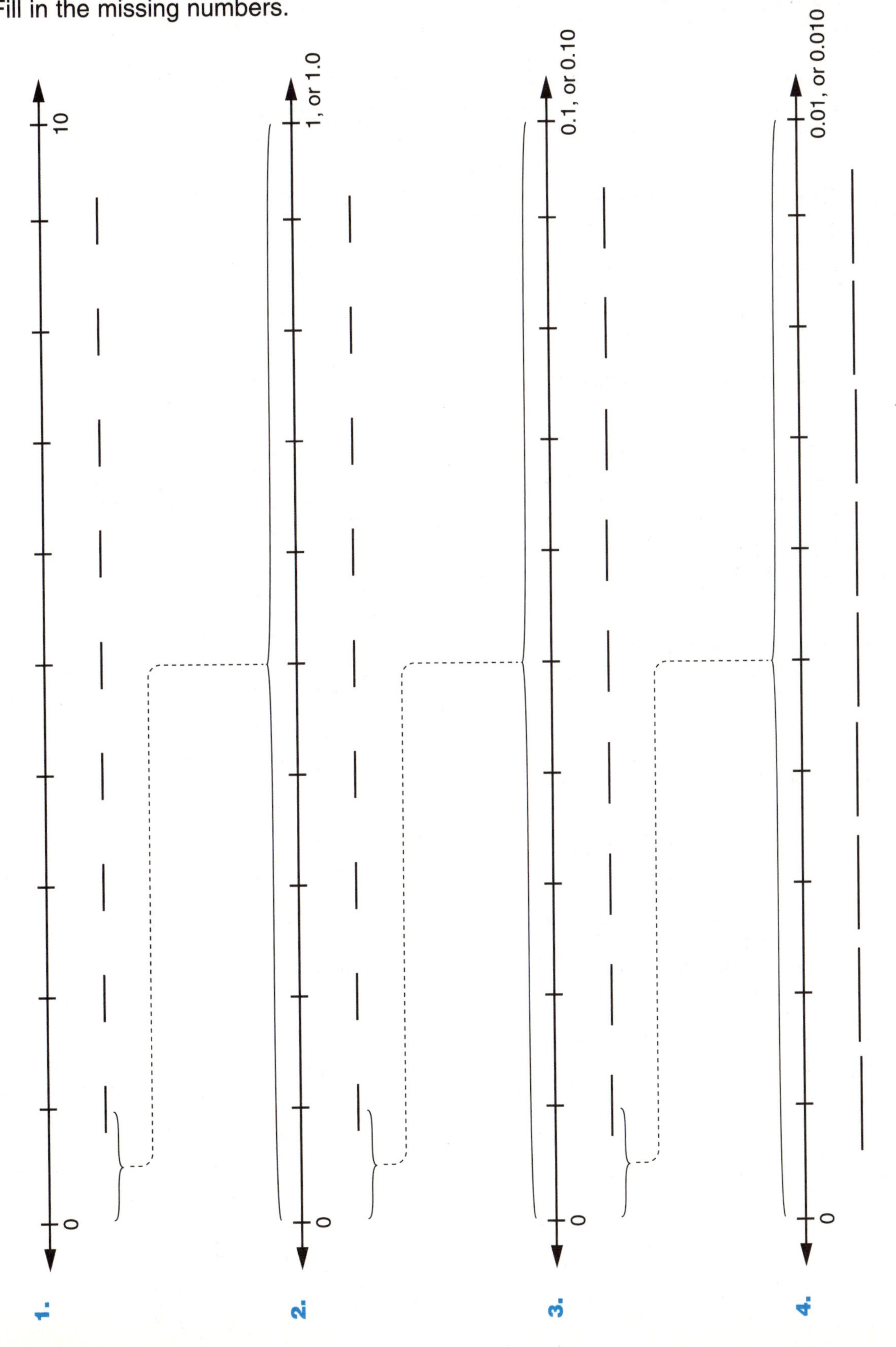

Use with Lesson 4.10.

Date Time

Broken-Calculator Problems

Solve the open sentences below with your calculator. Pretend one of the calculator keys is broken.

Example Pretend the ⊖ key is broken. What is the solution for the open sentence $275 + x = 921$?

Replace the variable x with a number and see if you get a true sentence. If it is not true, try other numbers until you get a true sentence. Here is one solution:

Try 700: $275 + \mathbf{700} = 975$	700 is too big.
Try 650: $275 + \mathbf{650} = 925$	Only 4 away from the solution.
Try 654: $275 + \mathbf{654} = 929$	Wrong way!
Try 646: $275 + \mathbf{646} = 921$	This is a true sentence. 646 is the solution.

1. Pretend the ⊖ key is broken. What is the solution for the open sentence $y + 481 = 1{,}106$? Solution: ________

2. Pretend the ⊕ key is broken. What is the solution for the open sentence $d - 396 = 235$? Solution: ________

3. Pretend the ⊕ key is broken. What is the solution for the open sentence $t * 52 = 3{,}536$? Solution: ________

4. Pretend the ⊖ key is broken. What is the solution for the open sentence $12.6 + m = 39.1$? Solution: ________

5. Pretend the ⊕ key is broken. What is the solution for the open sentence $30.46 - p = 13.72$? Solution: ________

6. Pretend the ⊕ key is broken. What is the solution for the open sentence $y - 20.2 = 75.13$? Solution: ________

Date Time

Math Boxes 4.10

1. a. Is 326 closer to 300 or 400?

   ____

   b. Name the number halfway between 500 and 800. ________

2. Without measuring, estimate the height of your chair. Then measure it.

   a. Estimate: About ________ cm

   b. Measurement: About ________ cm

3. Complete.

   a. 3 cm = ________ mm

   b. 15 cm = ________ mm

   c. ________ cm = 40 mm

   d. ________ cm = 3 m

   e. 500 cm = ________ m

   f. 2 m = ________ mm

4. Add or subtract.

   a. $\begin{array}{r} 647 \\ +\ 228 \\ \hline \end{array}$

   b. $\begin{array}{r} 500 \\ -\ 398 \\ \hline \end{array}$

5. Tell whether each number sentence is true or false.

   a. $2.34 - 0.09 = 2.25$ ________

   b. $89.6 + 21.7 = 111.3$ ________

   c. $56.4 - 23.8 < 33$ ________

   d. $5.17 + 3.86 > 10$ ________

6. Name two properties of a regular polygon.

   a. ________________________

   b. ________________________

Date Time

Time to Reflect

1. Apart from math class, describe two times when you have used decimals. Or, if you have not used decimals, describe two times when you could have used them.

2. What was the easiest lesson in this unit? Why was it easy?

3. What was the hardest lesson in this unit? What strategies did you use to try to understand it better?

Date Time

Math Boxes 4.11

MATH

1. Which of the following is closest to the sum of 721, 2,876, and 15,103?

   5,000

   10,000

   15,000

   20,000

2. Complete.

   a. $4 * 8 =$ ______

   b. $4 * 80 =$ ______

   c. ______ $= 5 * 3$

   d. ______ $= 50 * 3$

   e. $6 * 6 =$ ______

   f. $6 * 60 =$ ______

3. a. Is 63 closer to 60 or 70?

   ______

   b. Name the number halfway between 80 and 90. ______

   c. Is 572 closer to 500 or 600? ______

   d. Name the number halfway between 300 and 600. ______

4. Write the following numbers using digits:

   a. one million, three hundred forty-six thousand, thirteen

   ______

   b. twenty-two million, fifteen thousand, three hundred fifty-four

   ______

5. a.
$$\begin{array}{r} 35 \\ 100 \\ 280 \\ +\ 800 \\ \hline \end{array}$$

b.
$$\begin{array}{r} 18 \\ 420 \\ 120 \\ +\ 2{,}800 \\ \hline \end{array}$$

c.
$$\begin{array}{r} 54 \\ 180 \\ 360 \\ +\ 1{,}200 \\ \hline \end{array}$$

d.
$$\begin{array}{r} 48 \\ 720 \\ 180 \\ +\ 2{,}700 \\ \hline \end{array}$$

Multiplying Ones by Tens

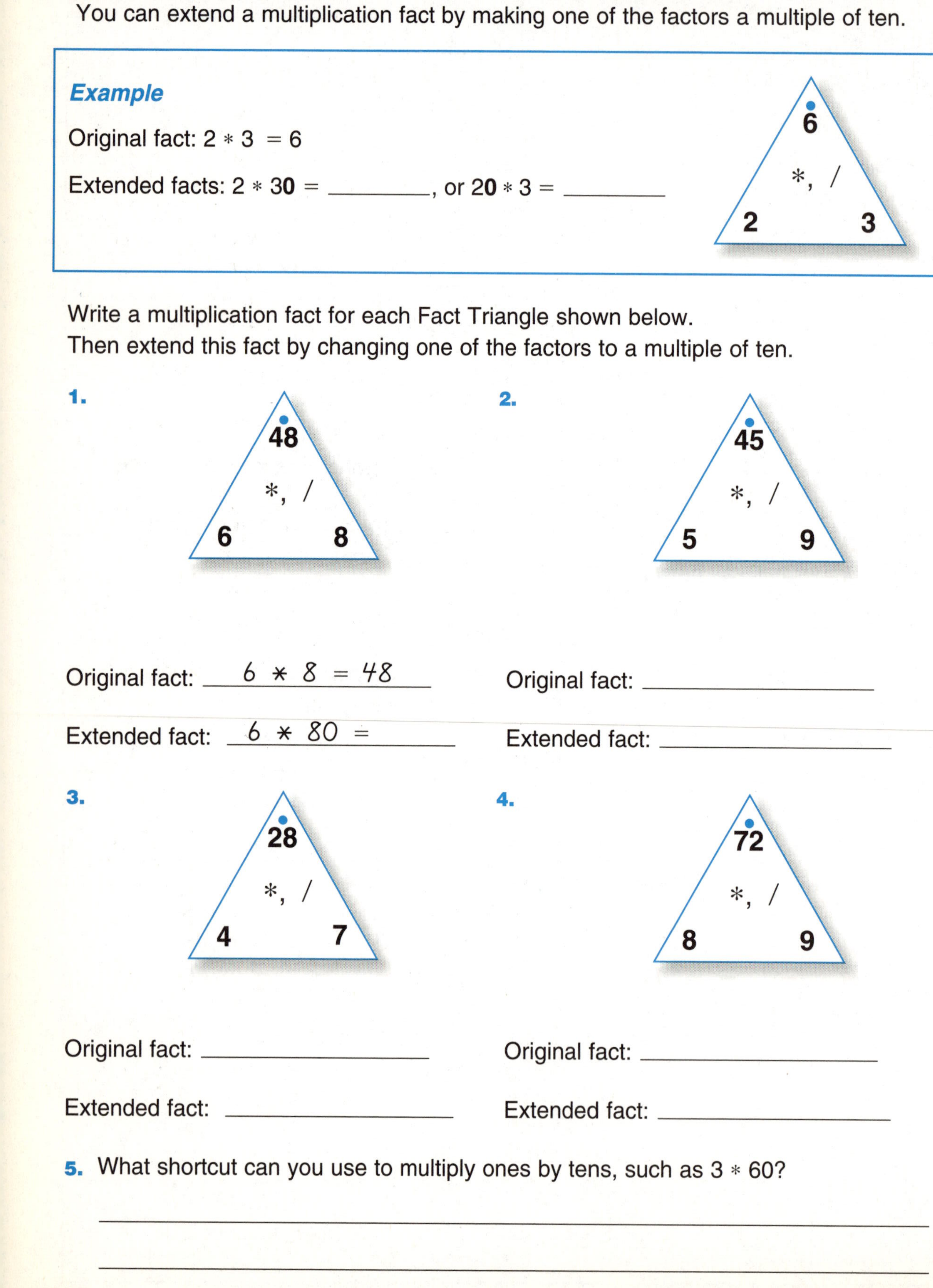

You can extend a multiplication fact by making one of the factors a multiple of ten.

Example

Original fact: 2 * 3 = 6

Extended facts: 2 * **30** = ________, or **20** * 3 = ________

Write a multiplication fact for each Fact Triangle shown below.
Then extend this fact by changing one of the factors to a multiple of ten.

1.

Original fact: 6 * 8 = 48

Extended fact: 6 * 80 = ________

2.

Original fact: ________

Extended fact: ________

3.

Original fact: ________

Extended fact: ________

4.

Original fact: ________

Extended fact: ________

5. What shortcut can you use to multiply ones by tens, such as 3 * 60?

____

____

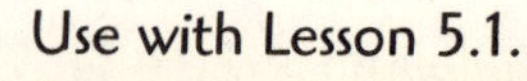

Date Time

Multiplying Tens by Tens

You can extend a multiplication fact by making both factors multiples of ten.

Example

Original fact: 3 * 5 = 15

Extended fact: **30** * **50** = ______

Write a multiplication fact for each Fact Triangle shown below.
Then extend this fact by changing both factors to multiples of ten.

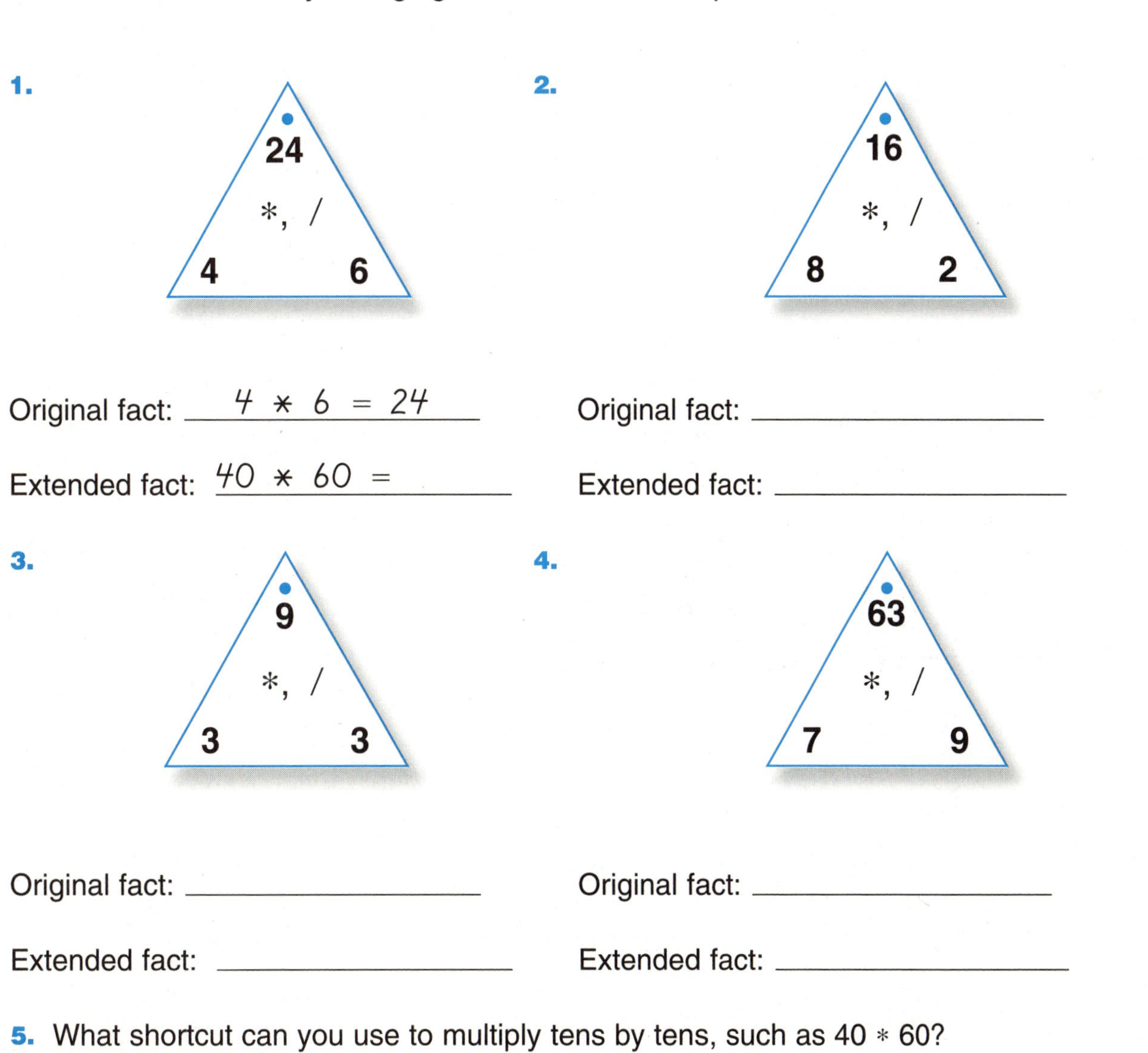

1.

Original fact: 4 * 6 = 24

Extended fact: 40 * 60 = ______

2.

Original fact: ______

Extended fact: ______

3.

Original fact: ______

Extended fact: ______

4.

Original fact: ______

Extended fact: ______

5. What shortcut can you use to multiply tens by tens, such as 40 * 60?

______

______

Date Time

Math Boxes 5.1

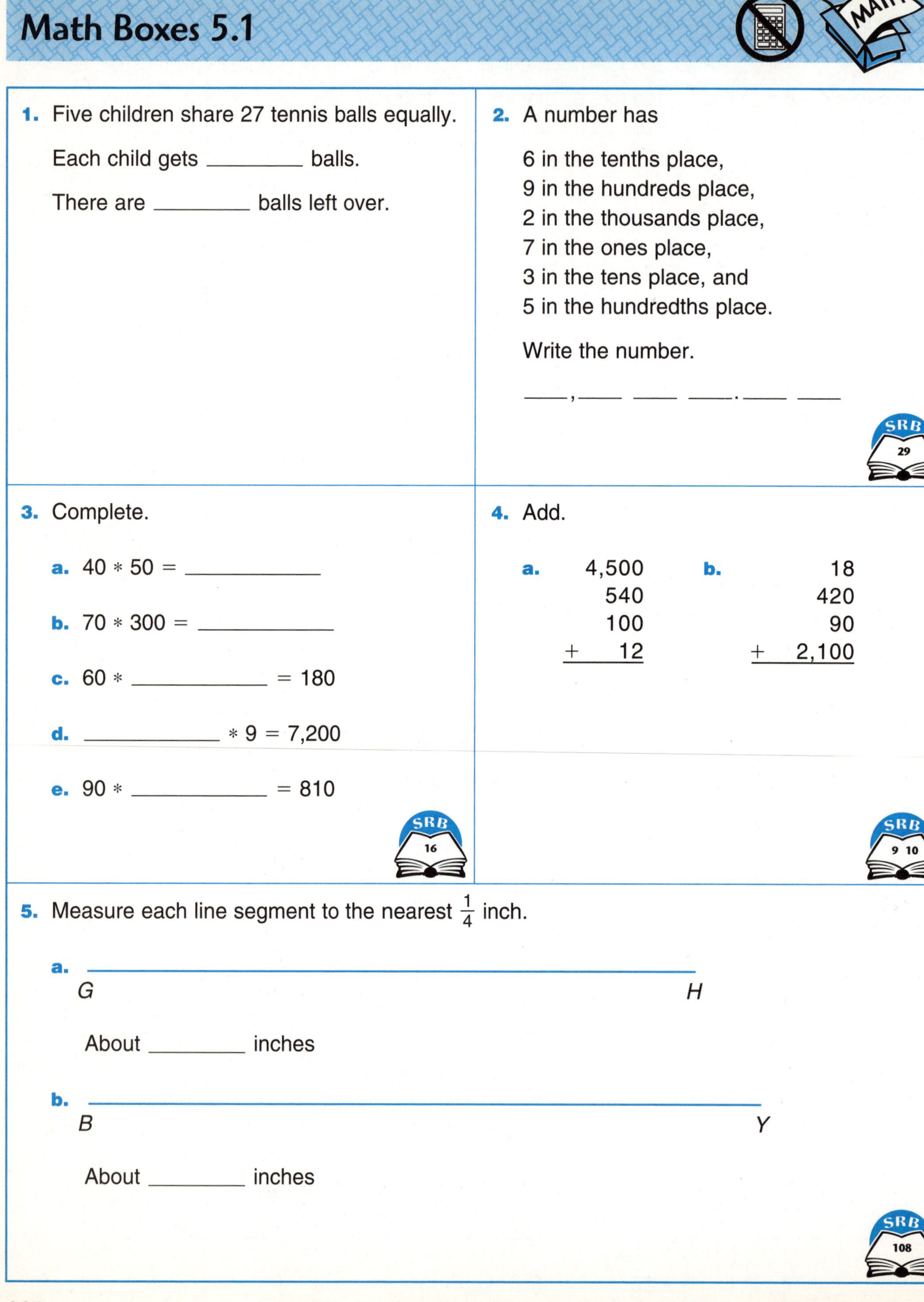

1. Five children share 27 tennis balls equally.

Each child gets ________ balls.

There are ________ balls left over.

SRB

2. A number has

6 in the tenths place,
9 in the hundreds place,
2 in the thousands place,
7 in the ones place,
3 in the tens place, and
5 in the hundredths place.

Write the number.

____, ____ ____ ____ . ____ ____

SRB 29

3. Complete.

a. 40 * 50 = ____________

b. 70 * 300 = ____________

c. 60 * ____________ = 180

d. ____________ * 9 = 7,200

e. 90 * ____________ = 810

SRB 16

4. Add.

a.
$$\begin{array}{r} 4{,}500 \\ 540 \\ 100 \\ +\ \ 12 \\ \hline \end{array}$$

b.
$$\begin{array}{r} 18 \\ 420 \\ 90 \\ +\ \ 2{,}100 \\ \hline \end{array}$$

SRB 9 10

5. Measure each line segment to the nearest $\frac{1}{4}$ inch.

a. G ______________________________ H

About ________ inches

b. B ________________________________ Y

About ________ inches

SRB 108

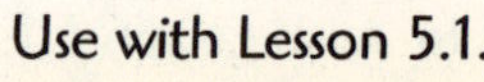

Date Time

Multiplication Wrestling Worksheet

Round 1 Cards: ______ ______ ______ ______

Numbers formed: ______ * ______

Teams: (______ + ______) * (______ + ______)

Products: ______ * ______ = ______

______ * ______ = ______

______ * ______ = ______

______ * ______ = ______

Total (add 4 products): ______

Round 2 Cards: ______ ______ ______ ______

Numbers formed: ______ * ______

Teams: (______ + ______) * (______ + ______)

Products: ______ * ______ = ______

______ * ______ = ______

______ * ______ = ______

______ * ______ = ______

Total (add 4 products): ______

Round 3 Cards: ______ ______ ______ ______

Numbers formed: ______ * ______

Teams: (______ + ______) * (______ + ______)

Products: ______ * ______ = ______

______ * ______ = ______

______ * ______ = ______

______ * ______ = ______

Total (add 4 products): ______

Date Time

Place Value in Decimals

1. Write these numbers in order from smallest to largest.

1.26 0.58 1.09 1.091 0.35

____

____

____

____

____

2. A number has

6 in the tenths place,
4 in the ones place,
5 in the hundredths place, and
9 in the tens place.

Write the number.

___ ___ . ___ ___

3. Write the smallest number you can make with the following digits:

3 6 4 7 2

____

4. What is the value of the digit 4 in the numerals below?

a. 37.**4**8 ________________

b. **4**9.08 ________________

c. 0.9**4**2 ________________

d. 1.66**4** ________________

5. Write each number using digits.

a. four and seventy-two hundredths

____

b. nine hundred thirty-five thousandths

____

6. I am a four-digit number less than 10.

- The digit in the tenths place is the result of dividing 36 by 4.
- The digit in the hundredths place is the result of dividing 42 by 7.
- The digit in the ones place is the result of dividing 72 by 8.
- The digit in the thousandths place is the result of dividing 35 by 5.

What number am I?

___ . ___ ___ ___

Date Time

Math Boxes 5.2

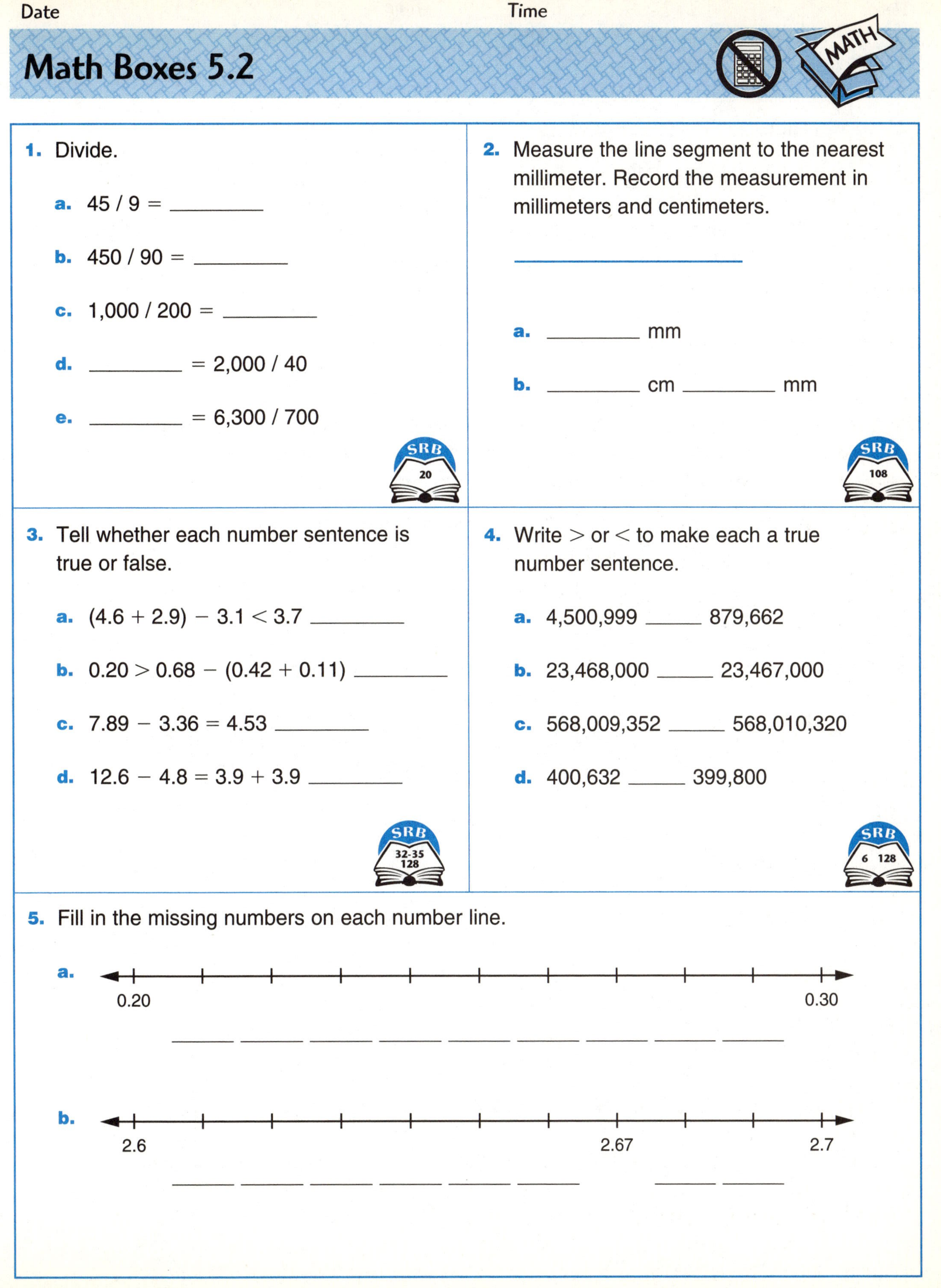

1. Divide.

   a. $45 / 9 =$ ______

   b. $450 / 90 =$ ______

   c. $1,000 / 200 =$ ______

   d. ______ $= 2,000 / 40$

   e. ______ $= 6,300 / 700$

SRB 20

2. Measure the line segment to the nearest millimeter. Record the measurement in millimeters and centimeters.

   a. ______ mm

   b. ______ cm ______ mm

SRB 108

3. Tell whether each number sentence is true or false.

   a. $(4.6 + 2.9) - 3.1 < 3.7$ ______

   b. $0.20 > 0.68 - (0.42 + 0.11)$ ______

   c. $7.89 - 3.36 = 4.53$ ______

   d. $12.6 - 4.8 = 3.9 + 3.9$ ______

SRB 32-35 128

4. Write > or < to make each a true number sentence.

   a. 4,500,999 ______ 879,662

   b. 23,468,000 ______ 23,467,000

   c. 568,009,352 ______ 568,010,320

   d. 400,632 ______ 399,800

SRB 6 128

5. Fill in the missing numbers on each number line.

   a. 0.20 ______ ______ ______ ______ ______ ______ ______ ______ ______ 0.30

   b. 2.6 ______ ______ ______ ______ ______ ______ 2.67 ______ ______ 2.7

Date Time

Planning a Driving Trip

Math Message

The airline you are using on the World Tour will give you a $200 discount coupon for every 15,000 miles you fly. Suppose you have flown the distances shown in the table at the right.

Route	Distance
Washington, D.C. → Cairo	5,980 mi
Cairo → Accra	2,420 mi
Accra → Cairo	2,420 mi
Cairo → Budapest	1,380 mi
Budapest → London	1,040 mi

Have you flown enough miles to get a discount coupon? ________

Use the map on journal page 119. Start at your hometown. Plan a driving trip that takes you to 4 other cities on the map. If your hometown is not on the map, find the nearest city on the map to your hometown. Start your driving trip from this city.

Example Trip Start in Chicago. Drive to St. Louis, Louisville, Birmingham, and then New Orleans.

You plan to drive about 8 hours a day and then stop for the night. You want to find out how many **days** this driving trip will take.

1. Record your routes, driving distances, and driving times in the table.

From...To	Driving Distance (miles)	Driving Time (hours:minutes)	Rounded Time (hours)
	___ ___ ___	___:___ ___	___
	___ ___ ___	___:___ ___	___
	___ ___ ___	___:___ ___	___
	___ ___ ___	___:___ ___	___

2. Estimate how many **miles** in all you will drive. About ________ miles

3. Estimate how many **hours** in all you will drive. About ________ hours

4. Tell how many days it will take to complete the trip. ________ days

Date Time

Estimated U.S. Distances and Driving Times

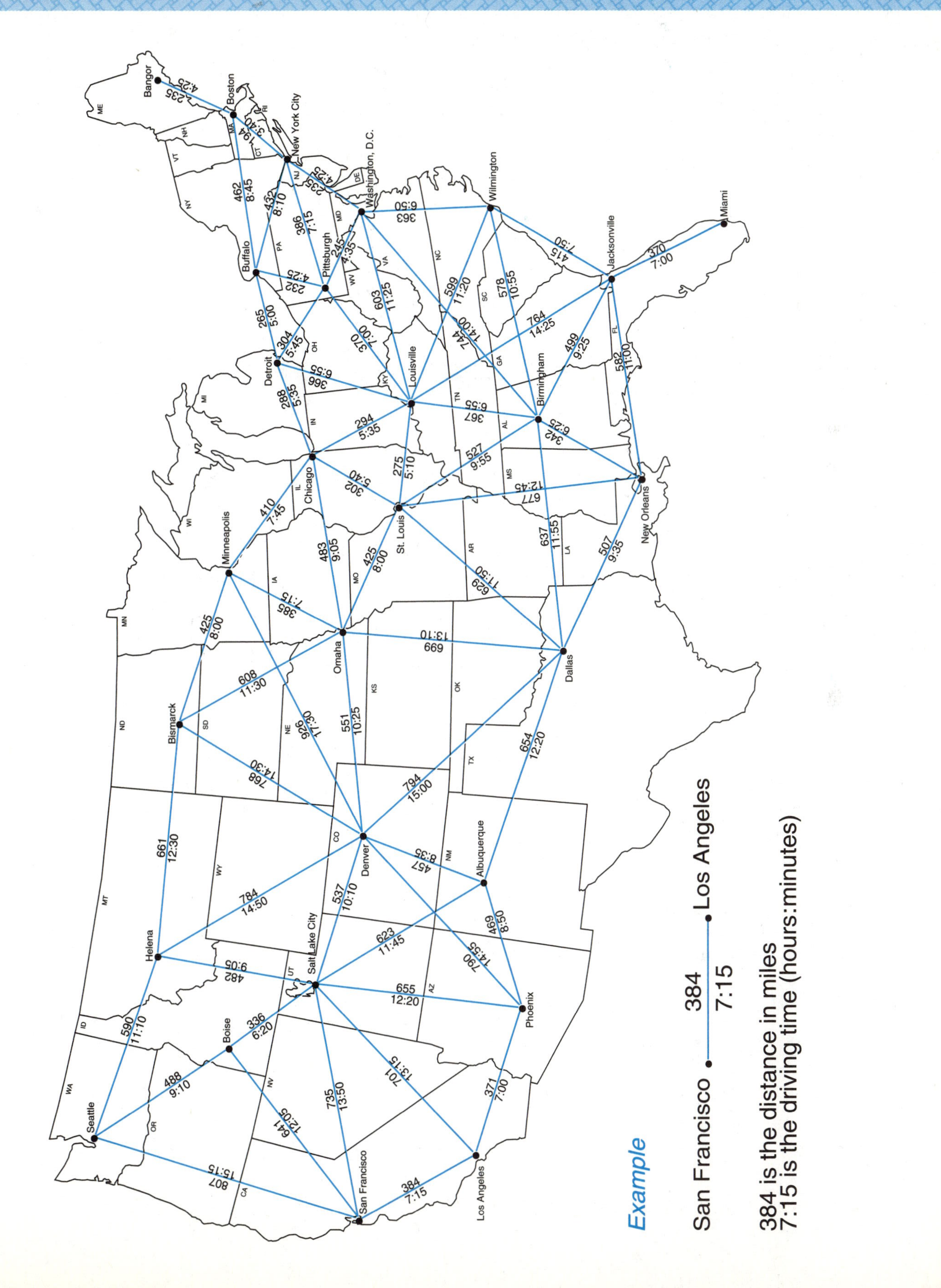

Date Time

Logic Problems

1. Theresa, Keiko, Jamal, and Mario live in Lansing, Michigan. Each has a favorite season. No two of them like the same season.

- Theresa likes to go to the beach.
- Keiko doesn't like snow.
- Jamal doesn't like spring flowers.
- Mario likes to go skiing.

What is each person's favorite season? You may use a logic grid to help you.

	Winter	Spring	Summer	Fall
Keiko				
Jamal				
Theresa				
Mario				

Keiko ________________ Jamal ________________

Theresa ________________ Mario ________________

2. Lynn, Julie, Elaine, and Pierre each have a favorite pizza. No two of them like the same topping. They ordered one of each of the following: shrimp, sausage, cheese, and spinach.

- Lynn is a vegetarian and will not eat meat or seafood.
- Pierre is allergic to seafood.
- Elaine likes meat toppings, but not seafood.
- Neither Pierre nor Julie likes vegetables.

What kind of pizza did each person eat?

Lynn ________________ Pierre ________________

Elaine ________________ Julie ________________

Date Time

Math Boxes 5.3

MATH

1. Sara collected 30 leaves. On the way to school, she lost 2 of them. At school, she shared them equally with her 6 friends. How many leaves did each friend get?

   ______ leaves

2. A number has

   2 in the hundreds place,
   7 in the tenths place,
   6 in the hundredths place,
   4 in the ones place,
   5 in the tens place, and
   1 in the thousandths place.

   Write the number.

   ___ ___ ___ . ___ ___ ___

3. Complete.

   a. 3 * 40 = ______

   b. 90 * 70 = ______

   c. 50 * ______ = 3,000

   d. ______ * 8 = 1,600

   e. 80 * ______ = 56,000

4. Add.

   a. 72 + 450 + 160 + 1,000

   b. 15 + 240 + 350 + 5,600

5. Measure each line segment to the nearest $\frac{1}{4}$ inch.

   a. E ______ R

   About ______ inches

   b. L ______ Y

   About ______ inches

Date Time

What Do Americans Eat?

Several years ago, the U.S. Department of Agriculture conducted a survey to find out how much food Americans eat. In the survey, a large number of people were asked to keep lists of all the foods they ate for several days.

These lists were then used to estimate how much of each food is eaten during one year. The "average" American eats about 1,431 pounds of food per year. This is about 4 pounds of food per day.

This survey found that the average American eats or drinks about the following amounts in one year:

2	pounds of broccoli
4	pounds of potato chips
7	pounds of peanuts
18	pounds of candy
27	pounds of lettuce
44	gallons of soft drinks
80	hot dogs
89	pounds of fresh fruit
121	pounds of potatoes
176	glasses of fruit juice
200	hamburgers
255	eggs
752	glasses of water

Use your answers to the Math Message question to complete these statements.

1. I will drink about ________ glasses of fruit juice in one year.

2. I will eat about ________ hot dogs in one year.

3. I will eat about ________ hamburgers in one year.

Date Time

Estimating Averages

- **Estimate** whether the answer will be in the tens, hundreds, thousands, or more.
- Write a number model to show how you estimated.
- Then circle the box that shows your estimate.

Example Alice sleeps an average of 9 hours a night. How many hours does she sleep in one year?

Number model $10 * 400 = 4,000$

10s	100s	1,000s (circled)	10,000s	100,000s	1,000,000s

1. An average of about 23 new species of insects are discovered each day. About how many new species are discovered in one year?

Number model ______________________

10s	100s	1,000s	10,000s	100,000s	1,000,000s

2. A housefly beats its wings about 190 times per second. That's about how many times per minute?

Number model ______________________

10s	100s	1,000s	10,000s	100,000s	1,000,000s

3. A blue whale weighs about as much as 425,000 kittens. About how many kittens weigh as much as 4 blue whales?

Number model ______________________

10s	100s	1,000s	10,000s	100,000s	1,000,000s

4. An average bee can lift about 300 times its own weight. If a 170-pound person were as strong as a bee, about how many pounds could this person lift?

Number model ______________________

10s	100s	1,000s	10,000s	100,000s	1,000,000s

Date Time

Math Boxes 5.4

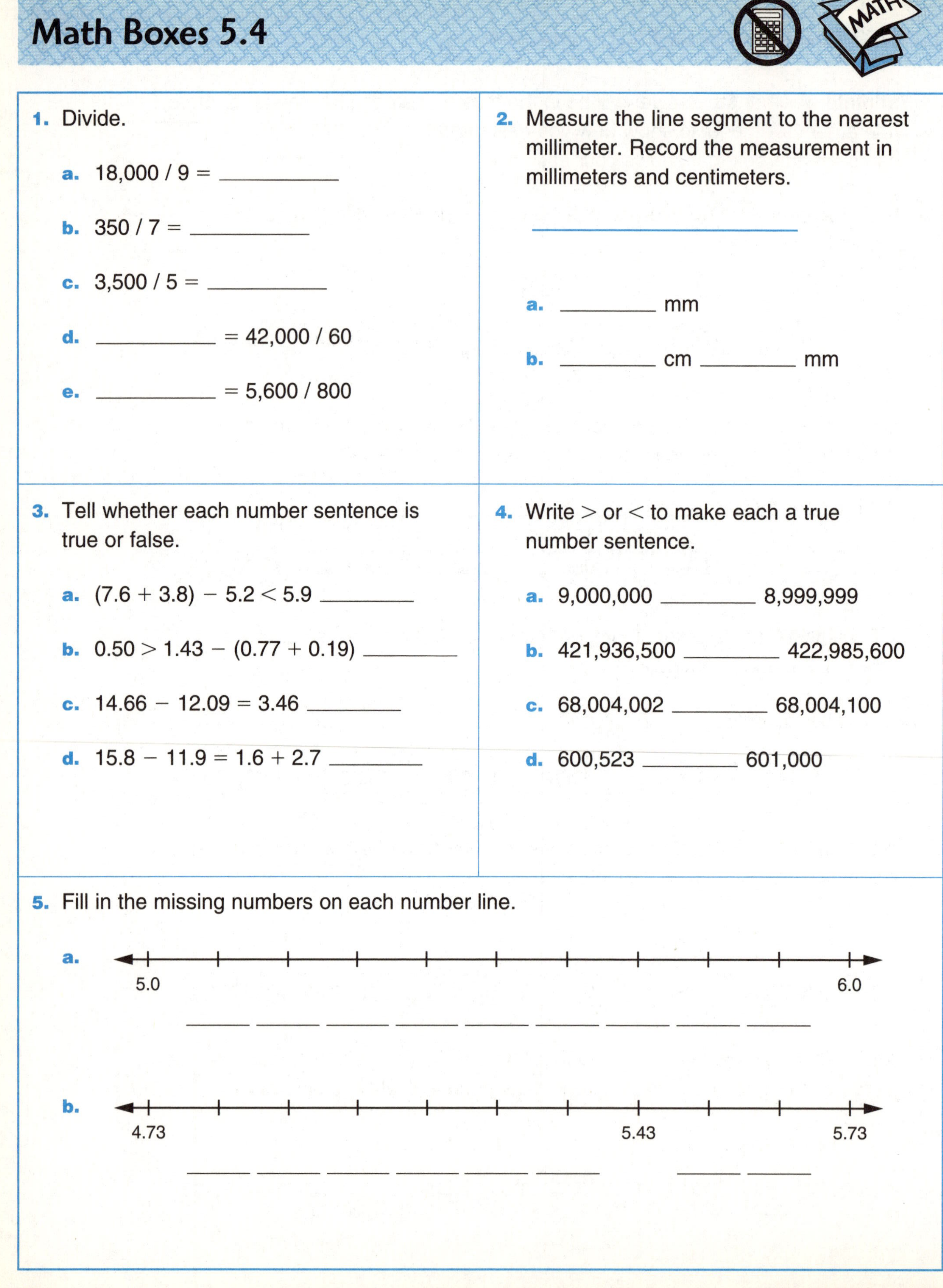

1. Divide.

a. $18{,}000 / 9 =$ ______

b. $350 / 7 =$ ______

c. $3{,}500 / 5 =$ ______

d. ______ $= 42{,}000 / 60$

e. ______ $= 5{,}600 / 800$

2. Measure the line segment to the nearest millimeter. Record the measurement in millimeters and centimeters.

a. ______ mm

b. ______ cm ______ mm

3. Tell whether each number sentence is true or false.

a. $(7.6 + 3.8) - 5.2 < 5.9$ ______

b. $0.50 > 1.43 - (0.77 + 0.19)$ ______

c. $14.66 - 12.09 = 3.46$ ______

d. $15.8 - 11.9 = 1.6 + 2.7$ ______

4. Write $>$ or $<$ to make each a true number sentence.

a. 9,000,000 ______ 8,999,999

b. 421,936,500 ______ 422,985,600

c. 68,004,002 ______ 68,004,100

d. 600,523 ______ 601,000

5. Fill in the missing numbers on each number line.

a. 5.0 ______ ______ ______ ______ ______ ______ ______ ______ ______ 6.0

b. 4.73 ______ ______ ______ ______ ______ ______ 5.43 ______ ______ 5.73

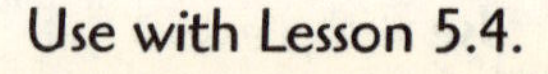

Date Time

The Partial-Products Method

1. Multiply. Show your work in the grid below.

   a. 59 * 6 = __________

   b. __________ = 3 * 470

   c. 2 * 1,523 = __________

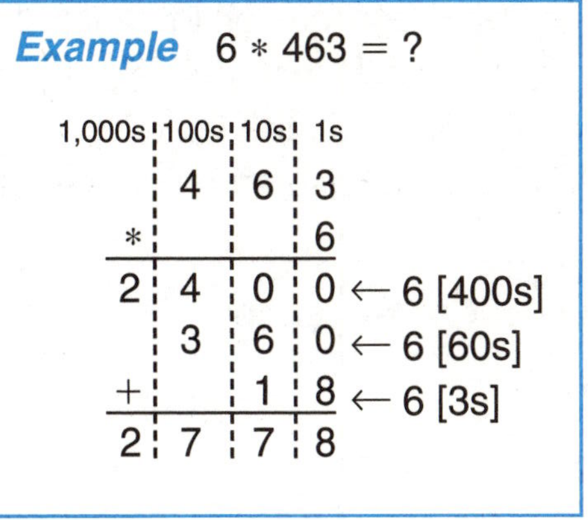

2. On an “average” day, about 217 pairs of twins are born in the United States. About how many pairs of twins are born in 1 week?

   a. **Estimate** whether the answer will be in the tens, hundreds, thousands, or more. Circle the box that shows your estimate.

10s	100s	1,000s	10,000s	100,000s	1,000,000s

   b. **Calculate** the answer in the grid below. About ________ pairs of twins

Date Time

My Measurement Collection for Units of Length

Use your personal references on journal page 100 to estimate the length or height of an object or distance in inches, feet, or yards. Describe the object or distance and record your estimate in the table below. Then measure the object or distance and record the actual measurement in the table.

Object or Distance	Estimated Length	Actual Length

Date Time

Math Boxes 5.5

1. Look at the grid below.

	A	B	C
1			
2			○
3			

a. In which column is the circle located?

____

b. In which row is the circle located?

____

2. Write each number using digits.

a. five million, two hundred sixty-eight thousand, four

____

b. two hundred million, three thousand, eighty-eight

____

SRB 4

3. ~~Circle~~ Write the number that is closest to the sum of 387, 945, and 1,024.

1,600

2,000

2,400

2,800

SRB 155 156

4. Multiply. ~~Use the partial-products method~~.

9 ∗ 237 = _______

SRB 17

5. Carlos, Frank, Mia, and Lauren each have a different favorite fruit: apples, grapes, watermelon, or peaches.

- Carlos likes a fruit that has a large seed in it.
- Mia likes a smaller fruit that is usually green or red.
- Lauren does not like watermelon or grapes.

Which fruit does each student like best?

Lauren _______________ Mia _______________

Carlos _______________ Frank _______________

SRB 149

Date ______ Time ______

Multiplication Number Stories

Follow these steps for each problem.

a. Decide which two numbers need to be multiplied to give the exact answer. Write the two numbers.

b. Estimate whether the answer will be in the tens, hundreds, thousands, or more. Write a number model for the estimate. Circle the box to show your estimate.

c. On the grid below, find the exact answer by multiplying the two numbers. Write the answer.

1. The average person drinks about 15 glasses of milk a month. About how many glasses of milk is that per year?

a. ______ * ______
numbers that will give the exact answer

b. ____________________
number model for your estimate

c. __________
exact answer

10s	100s	1,000s	10,000s	100,000s	1,000,000s

2. Eighteen newborn hummingbirds weigh about 1 ounce. About how many of them does it take to make 1 pound? (1 pound = 16 ounces)

a. ______ * ______
numbers that will give the exact answer

b. ____________________
number model for your estimate

c. __________
exact answer

10s	100s	1,000s	10,000s	100,000s	1,000,000s

Date Time

Multiplication Number Stories (cont.)

3. A test found that a lightbulb lasts an average of 63 days after being turned on. About how many hours is that?

a. ______ * ______
numbers that will give the exact answer

b. ____________________
number model for your estimate

c. __________
exact answer

10s	100s	1,000s	10,000s	100,000s	1,000,000s

4. A full-grown oak tree loses an average of about 7 tons of water through its leaves per day. About how many tons of water is that per year?

a. ______ * ______
numbers that will give the exact answer

b. ____________________
number model for your estimate

c. __________
exact answer

10s	100s	1,000s	10,000s	100,000s	1,000,000s

Date Time

Math Boxes 5.6

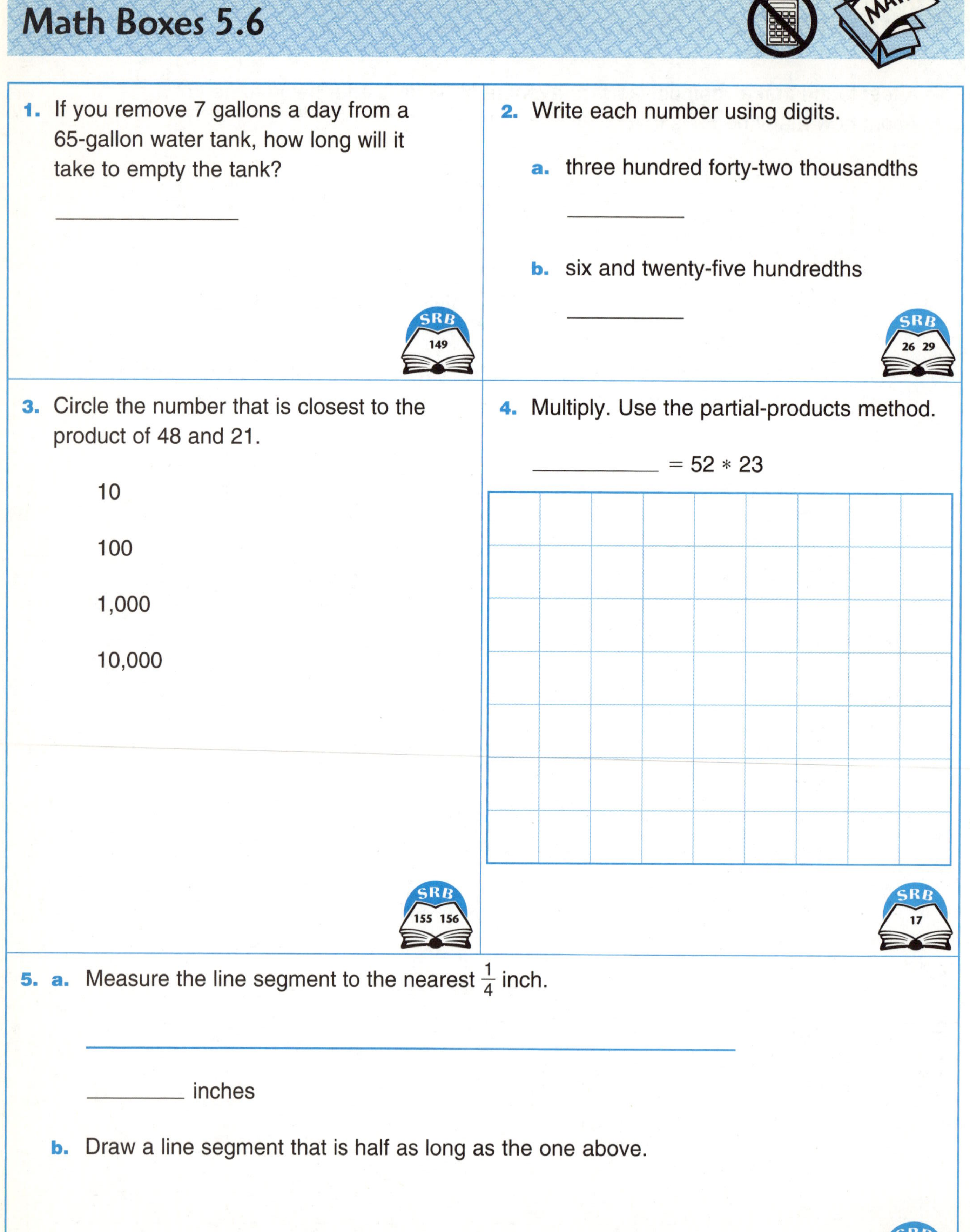

1. If you remove 7 gallons a day from a 65-gallon water tank, how long will it take to empty the tank?

____

SRB 149

2. Write each number using digits.

a. three hundred forty-two thousandths

____

b. six and twenty-five hundredths

____

SRB 26 29

3. Circle the number that is closest to the product of 48 and 21.

10

100

1,000

10,000

SRB 155 156

4. Multiply. Use the partial-products method.

________ $= 52 * 23$

SRB 17

5. a. Measure the line segment to the nearest $\frac{1}{4}$ inch.

________ inches

b. Draw a line segment that is half as long as the one above.

c. How long is the line segment you drew? ________ inches

SRB 108

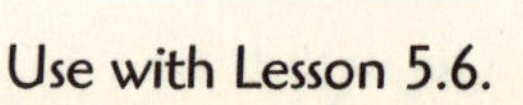

Date Time

Lattice Multiplication

Part A Use the lattice method to find the products.

1. 3 * 56 = ________

5 6 / 3

2. 8 * 26 = ________

2 6 / 8

3. 7 * 74 = ________

4. 6 * 315 = ________

5. 9 * 284 = ________

Part B Use the lattice method to find the products.

6. 47 * 63 = ________

6 3 / 4 / 7

7. 26 * 26 = ________

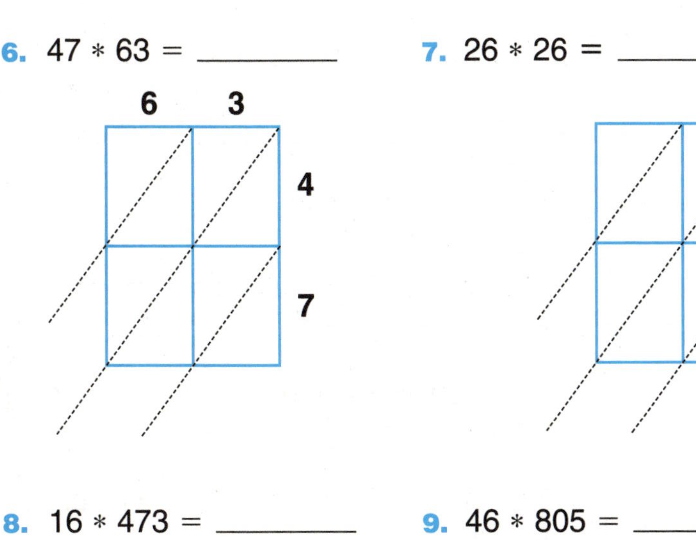

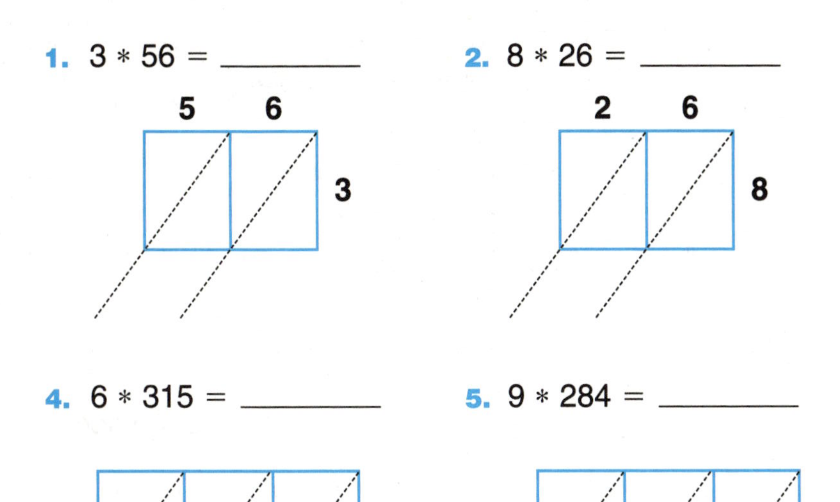

8. 16 * 473 = ________

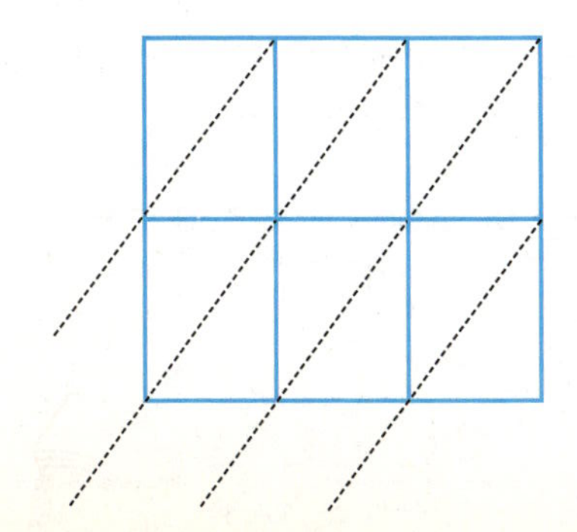

9. 46 * 805 = ________

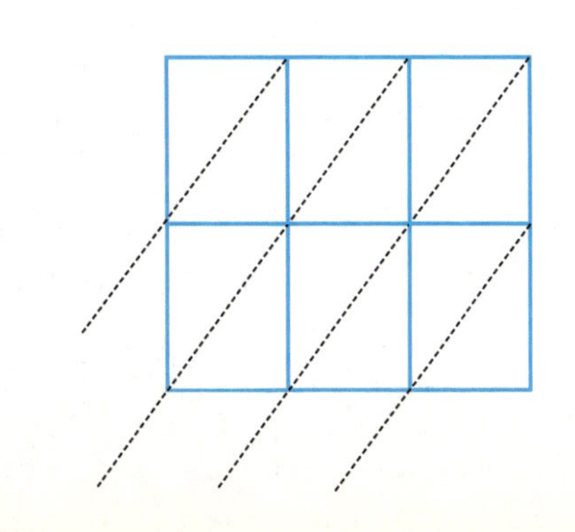

Date Time

Place Value in Whole Numbers

1. Write these numbers in order from smallest to largest.

964 9,460 96,400 400,960 94,600

____

____

____

____

____

2. A number has

5 in the hundreds place,
7 in the ten-thousands place,
0 in the ones place,
9 in the thousands place, and
8 in the tens place.

Write the number.

___ ___, ___ ___ ___

3. Write the largest number you can make with the following digits:

3 5 0 7 9 2

____

4. What is the value of the digit 8 in the numerals below?

a. **8**07,941 ________________

b. 5**8**3 ________________

c. **8**,714 ________________

d. **8**6,490 ________________

5. Write each number using digits.

a. four hundred eighty-seven thousand, sixty-three

____

b. fifteen thousand, two hundred ninety-seven

____

6. I am a 5-digit number.

- The digit in the thousands place is the result of dividing 64 by 8.
- The digit in the ones place is the result of dividing 63 by 9.
- The digit in the ten-thousands place is the result of dividing 54 by 6.
- The digit in the tens place is the result of dividing 40 by 5.
- The digit in the hundreds place is the result of dividing 33 by 11.

What number am I?

___ ___, ___ ___ ___

Use with Lesson 5.7.

Date Time

Math Boxes 5.7

1. Look at the grid below.

a. In which column is the triangle located?

______

b. In which row is the triangle located?

______

	A	B	C
1			
2			
3		△	

2. Write each number using digits.

a. twenty-six million, nineteen thousand, eighteen

______

b. three hundred fifty-two million, eight hundred thousand, two hundred

______

3. Circle the number that is closest to the sum of 715, 1,904, and 688.

3,000

3,300

3,600

3,900

4. Multiply. Use the partial-products method.

$6 * 509 =$ ______

5. Matthew, Mark, Elizabeth, and John each have a different favorite kind of book: mystery, science fiction, biography, or historical fiction.

- Matthew likes books with lots of clues.
- John likes books about events that might take place in the future.
- Elizabeth does not like to read about anything that didn't really happen.

Which kind of book does each student like best?

Matthew ______ Elizabeth ______

Mark ______ John ______

Date Time

Reading and Writing Big Numbers

1. Each row in the place-value chart shows a number. Write the name for each number below the chart.

	Billions				Millions				Thousands				Ones		
	100B	10B	1B	,	100M	10M	1M	,	100Th	10Th	1Th	,	100	10	1
a.			7	,	4	0	0	,	0	6	5	,	2	0	0
b.						5	1	,	8	0	0	,	0	0	0
c.		2	3	,	0	0	0	,	0	0	5	,	1	4	0
d.	1	2	3	,	4	5	6	,	7	8	9	,	0	1	2

a. 7 billion, 400 million, 65 thousand, 200

b. ______________________

c. ______________________

d. ______________________

2. Write these numbers in the place-value chart below.

a. 400 thousand, 500

b. 208 million, 350 thousand, 600

c. 16 billion, 210 million, 48 thousand, 715

d. 1 billion, 1 million, 1 thousand, 1

	Billions				Millions				Thousands				Ones		
	100B	10B	1B	,	100M	10M	1M	,	100Th	10Th	1Th	,	100	10	1
a.															
b.															
c.															
d.															

Date Time

How Much Are a Million and a Billion?

1. How many dots are on the 50 by 40 array page? ________ dots

2. How many dots would be on

a. 5 pages? ________ dots

b. 50 pages? ________ dots

c. 500 pages? ________ dots

3. Each package of paper, or ream, contains 500 sheets. How many dots would be on the paper in

a. 1 ream? (*Hint:* Look at Problem 2.) ________ dots

b. 10 reams? (1 carton) ________ dots

c. 100 reams? (10 cartons) ________ dots

d. 1,000 reams? (100 cartons) ________ dots

4. Write these numbers in the place-value chart below.

a. 999 thousand **b.** 1,000 thousand **c.** 999 million **d.** 1,000 million

	Billions				Millions				Thousands				Ones		
	100B	10B	1B	,	100M	10M	1M	,	100Th	10Th	1Th	,	100	10	1
a.															
b.															
c.															
d.															

Date Time

Math Boxes 5.8

1. You have 75 cookies. You give 8 cookies to each of your friends until you run out of cookies.

   a. How many friends will get 8 cookies? ______

   b. How many cookies will you get if you keep the rest for yourself? ______

2. Write each number using digits.

   a. seven hundred six thousandths

   ______

   b. three and four hundredths

   ______

3. Circle the number that is closest to the product of 41 and 83.

   30

   300

   3,000

   30,000

4. Multiply. Use the partial-products method.

   ______ = 46 * 98

5. a. Measure the line segment to the nearest $\frac{1}{4}$ inch.

   R S

   ______ inches

   b. Draw a line segment that is half the length of $\overline{RS}$.

   c. How long is the line segment you drew? ______ inches

Date

Time

Place Value and Powers of 10

	Millions	Hundred-Thousands	Ten-Thousands	Thousands	Hundreds	Tens	Ones
1.	1,000,000			1,000	100		1
2.	10 [100,000s]			10 [100s]	10 [10s]		10 [tenths]
3.			10 * 10 * 10 * 10		10 * 10		
4.		10^5		10^3	10^2		10^0

Fill in this place-value chart as follows:

1. Write standard numbers in Row 1.
2. In Row 2, write the value of each place to show that it is 10 times the value of the place to its right.
3. In Row 3, write the place values as products of 10s.
4. In Row 4, show the values as powers of 10. Use exponents. The exponent shows how many times 10 is used as a factor. It also shows how many zeros are in the standard number.

Date Time

Multiplication of Whole Numbers

Use the **partial-products** method to solve Problems 1 and 2.

1. The school lunchroom purchased 55 dozen eggs. How many eggs were purchased?

   ________ eggs

2. 76 * 59 = ________

Use the **lattice method** to solve Problems 3 and 4.

3. 23 * 314 = ________

4. An opossum sleeps an average of 19 hours per day. How many hours does an opossum sleep during a 4-week time period?

   ________ hours

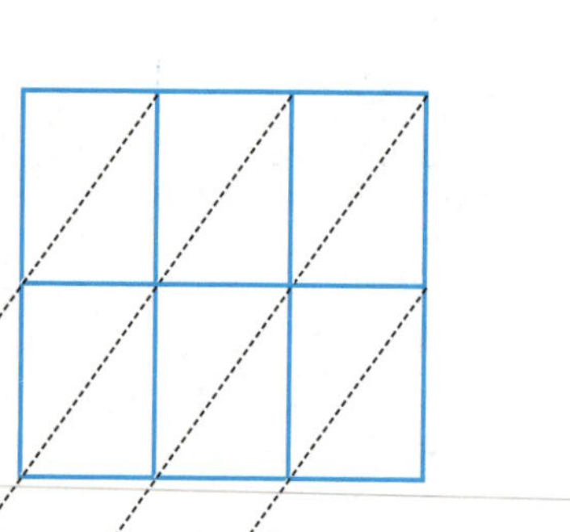

5. Write a multiplication number story using the numbers 85 and 64. You may use the method of your choice to solve it.

____

____

____

____

Answer: ________________
(unit)

Date Time

Math Boxes 5.9

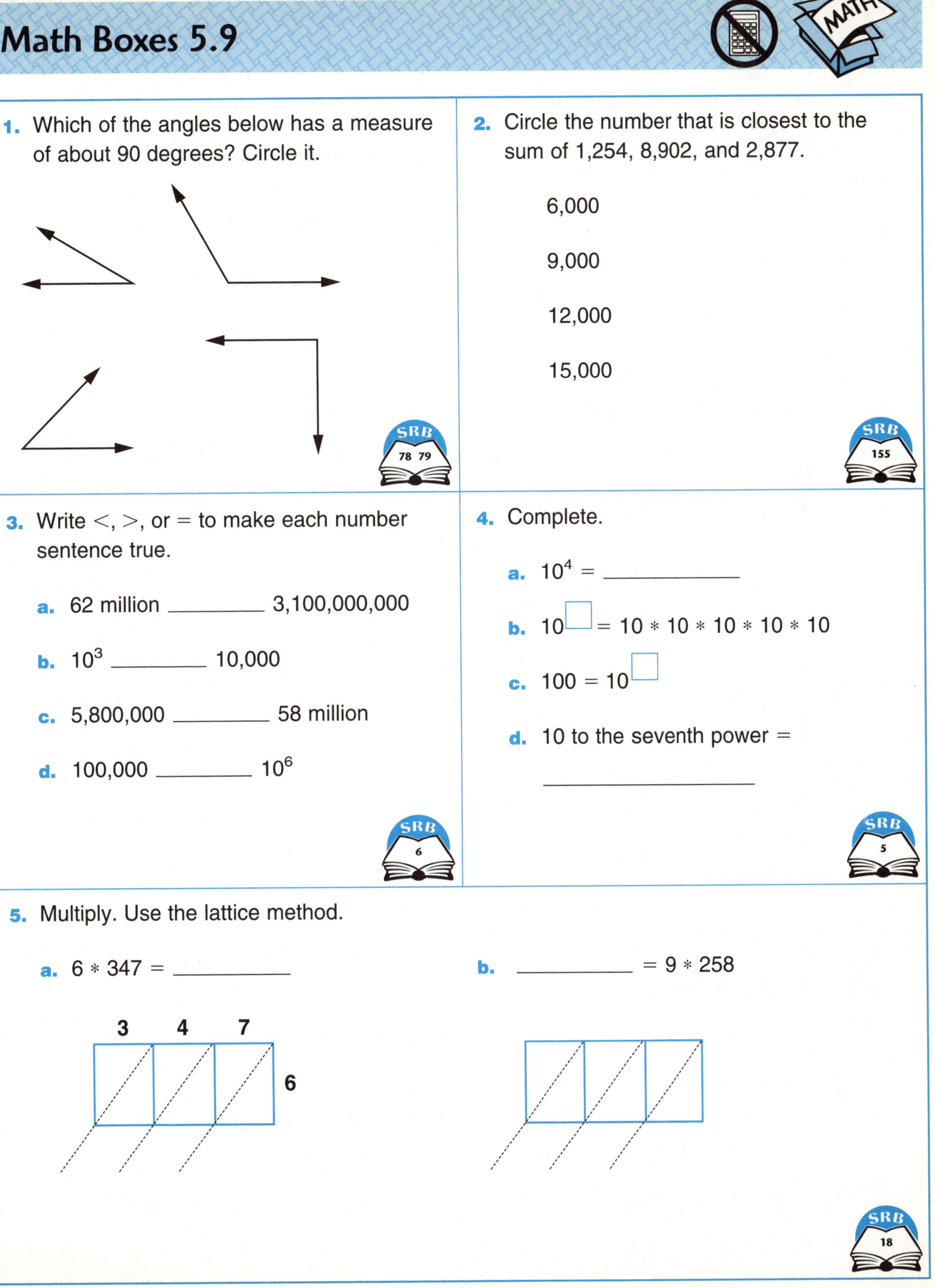

1. Which of the angles below has a measure of about 90 degrees? Circle it.

SRB 78 79

2. Circle the number that is closest to the sum of 1,254, 8,902, and 2,877.

6,000

9,000

12,000

15,000

SRB 155

3. Write <, >, or = to make each number sentence true.

a. 62 million ________ 3,100,000,000

b. 10^3 ________ 10,000

c. 5,800,000 ________ 58 million

d. 100,000 ________ 10^6

SRB 6

4. Complete.

a. 10^4 = ________

b. $10^{\square}$ = 10 * 10 * 10 * 10 * 10

c. 100 = $10^{\square}$

d. 10 to the seventh power =

____

SRB 5

5. Multiply. Use the lattice method.

a. 6 * 347 = ________

3 4 7

6

b. ________ = 9 * 258

SRB 18

Date Time

Evaluating Large Numbers

1. Round the attendance figures in the table to the nearest hundred-thousand.

1999 Major League Baseball—Attendance of Home Games for 10 Teams

Team	Attendance	Attendance Rounded to the Nearest 100,000
1. Colorado Rockies	3,481,065	
2. Baltimore Orioles	3,432,099	
3. Cleveland Indians	3,468,436	
4. Los Angeles Dodgers	3,098,042	
5. Atlanta Braves	3,284,901	
6. Texas Rangers	2,774,501	
7. Seattle Mariners	2,915,908	
8. St. Louis Cardinals	3,235,833	
9. Boston Red Sox	2,446,277	
10. New York Yankees	3,293,659	

Source: Information Please Sports Almanac

2. How do you think attendance figures for major league baseball games are obtained?

____

____

3. Do you think *exactly* 3,481,065 people were at the home games played by the Colorado Rockies? Explain your answer.

____

____

4. You rounded the figures in the table above to the nearest hundred-thousand. Which two pairs of teams have the same attendance figures based on these rounded numbers?

____

____

Date Time

Many Names for Powers of 10

At the bottom of the page, there are different names for powers of 10.
Write each of these names in the appropriate name-collection box.
Compare your answers with your partner's.

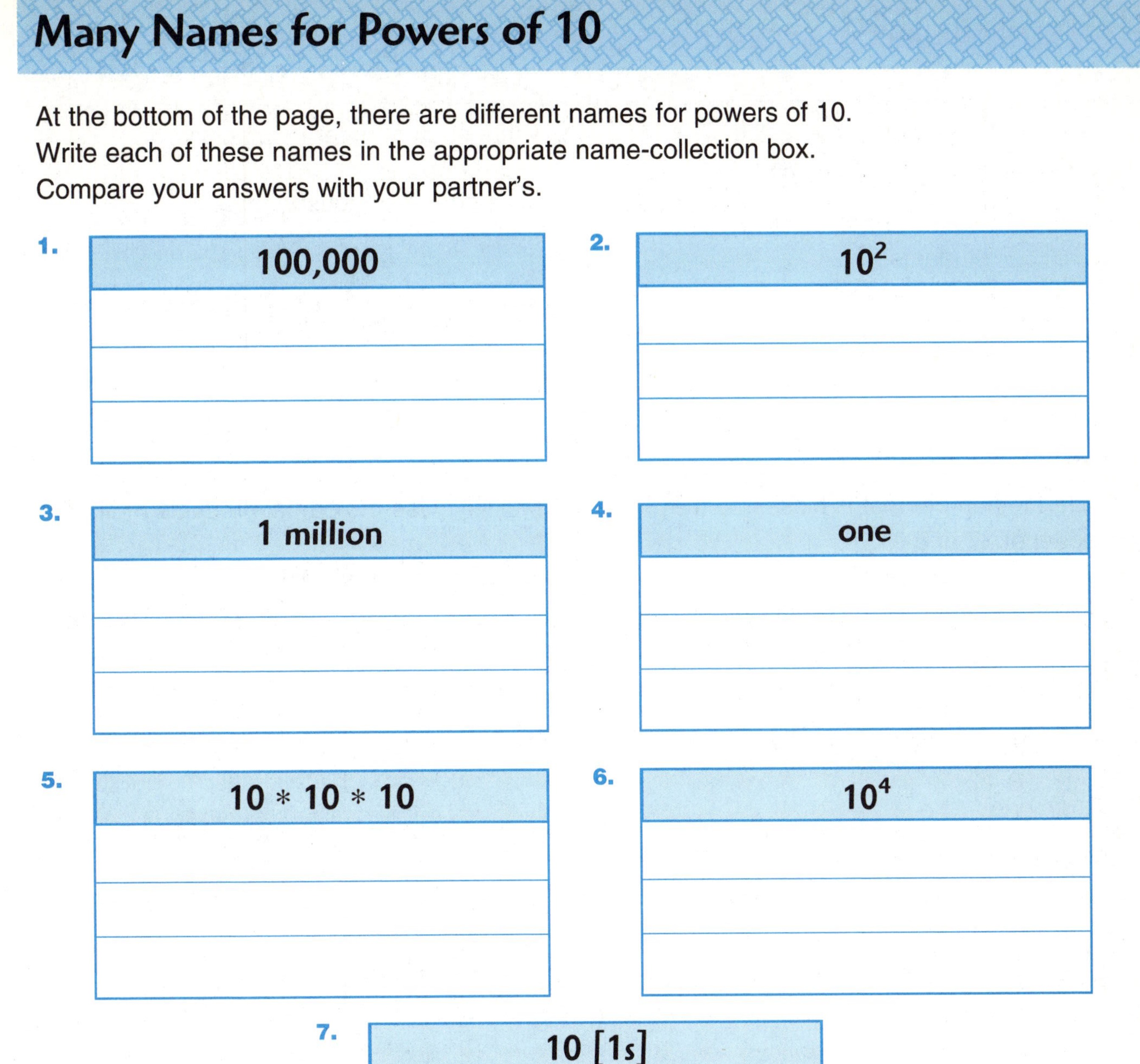

1. **100,000**

2. **10^2**

3. **1 million**

4. **one**

5. **10 * 10 * 10**

6. **10^4**

7. **10 [1s]**

1,000,000	10,000	1,000
100	10	10 [100,000s]
10 [10,000s]	10^6	10 [1,000s]
10^3	10 * 10 * 10 * 10	one thousand
10^5	10 * 10 * 10 * 10 * 10	10 [10s]
10 * 10	ten	10^1
10 [tenths]	10^0	1

Use with Lesson 5.10.

Date Time

Math Boxes 5.10

1. There are 60 trading cards. Each student gets 5 cards. How many students are there?

________ students

2. Write each number using digits.

a. eighty-five thousandths

____

b. fourteen and three tenths

____

3. Circle the number that is closest to the product of 37 and 91.

36

360

3,600

36,000

4. Multiply. Use the partial-products method.

________ $= 43 * 89$

5. a. Measure the line segment to the nearest $\frac{1}{4}$ inch.

T G

About ________ inches

b. Draw a line segment that is half the length of $\overline{TG}$.

c. How long is the line segment you drew? ________ inches

Date Time

Traveling to Europe

It is time to leave Africa. Your destination is Region 2—the continent of Europe. You and your classmates will fly from Cairo, Egypt to Budapest, Hungary. Before exploring Hungary, you will collect information about the countries in Region 2. You may even decide to visit another country in Europe after your stay in Budapest.

Use the World Tour section of your *Student Reference Book* to answer the questions. Use the Country Profiles for Region 2 on page 224.

1. Which country in Region 2 has

a. the largest population? ______________________
country population

b. the smallest population? ______________________
country population

c. the largest area? ______________________
country area

d. the smallest area? ______________________
country area

Use the Climate and Elevation of Capital Cities table on page 241.

2. From December to February, which capital in Region 2 has

a. the warmest weather? ______________________
capital country temperatures

b. the coolest weather? ______________________
capital country temperatures

c. the greatest amount of rain? ______________________
capital country inches rainfall

d. the least amount of rain? ______________________
capital country inches rainfall

Use the Population Data table on page 245.

3. Which country in Region 2 has

a. the greatest percent of population ages 0–14? ______________________
country percent

b. the smallest percent of population ages 0–14? ______________________
country percent

Date Time

Math Boxes 5.11

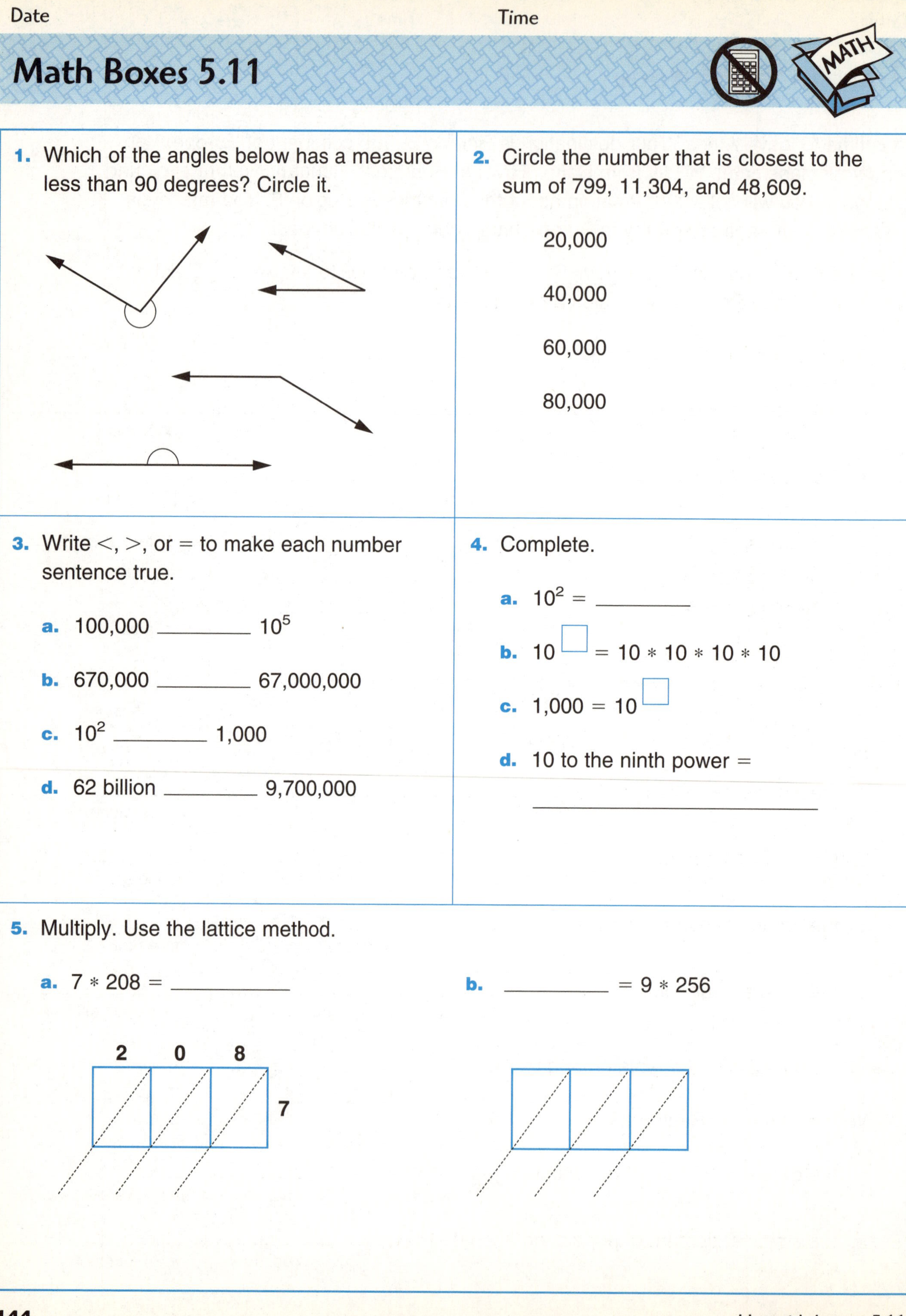

1. Which of the angles below has a measure less than 90 degrees? Circle it.

2. Circle the number that is closest to the sum of 799, 11,304, and 48,609.

20,000

40,000

60,000

80,000

3. Write <, >, or = to make each number sentence true.

a. 100,000 ______ 10^5

b. 670,000 ______ 67,000,000

c. 10^2 ______ 1,000

d. 62 billion ______ 9,700,000

4. Complete.

a. 10^2 = ______

b. $10^{\square}$ = 10 * 10 * 10 * 10

c. 1,000 = $10^{\square}$

d. 10 to the ninth power =

______

5. Multiply. Use the lattice method.

a. 7 * 208 = ______

2 0 8

7

b. ______ = 9 * 256

Date Time

Time to Reflect

1. During the past few weeks, about how many times did you use estimation outside of school? Describe some of those situations. Or, if you did not estimate, describe two times when you could have estimated.

2. Which of the big number names is the hardest for you to remember? Invent a memory device or method to help you remember how many zeros go after this number. Explain your trick.

3. Do you think multiplication is easy or hard? Explain your answer.

Date Time

Math Boxes 5.12

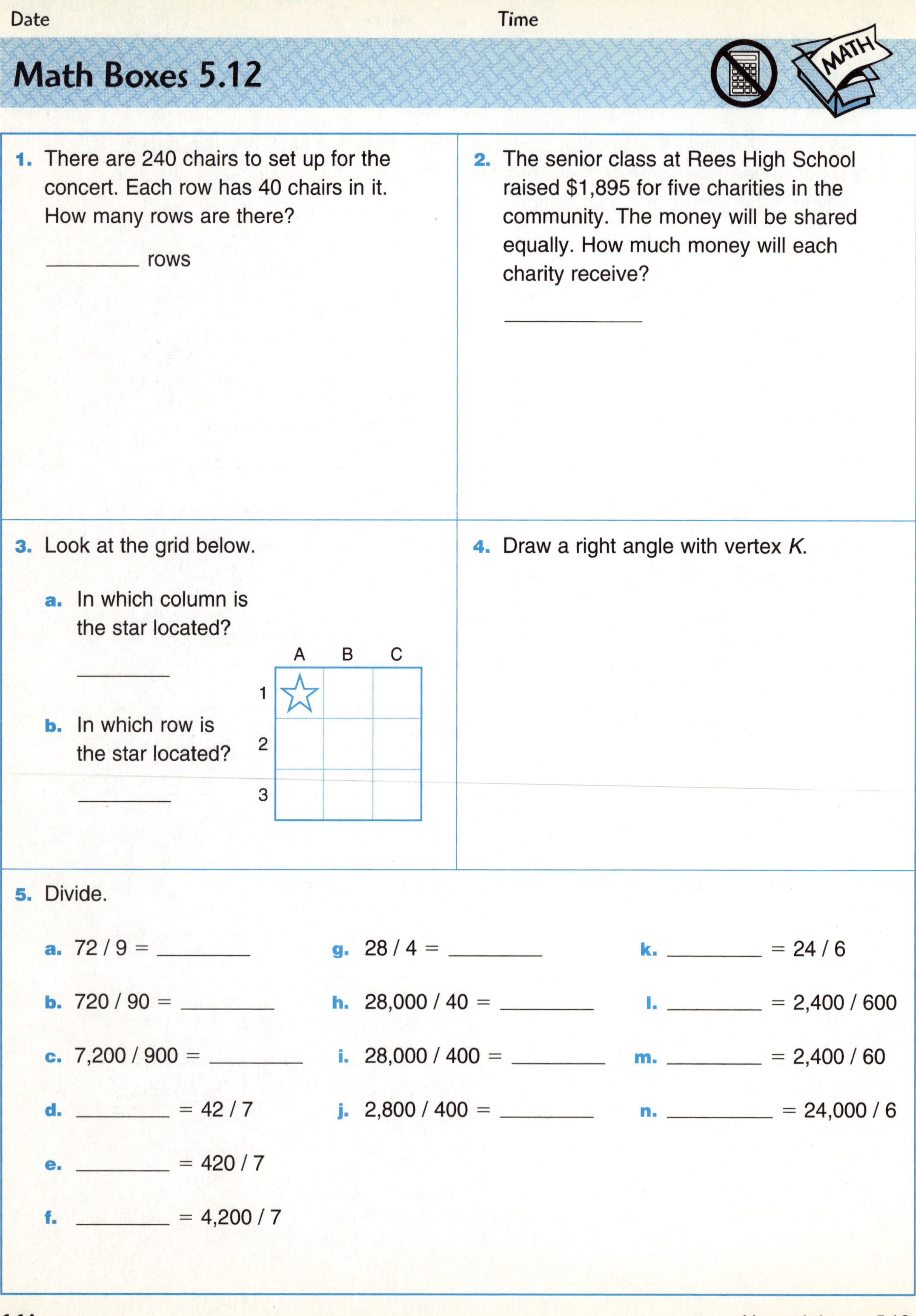

1. There are 240 chairs to set up for the concert. Each row has 40 chairs in it. How many rows are there?

_______ rows

2. The senior class at Rees High School raised $1,895 for five charities in the community. The money will be shared equally. How much money will each charity receive?

____

3. Look at the grid below.

a. In which column is the star located?

____

b. In which row is the star located?

____

4. Draw a right angle with vertex *K*.

5. Divide.

a. 72 / 9 = _______

b. 720 / 90 = _______

c. 7,200 / 900 = _______

d. _______ = 42 / 7

e. _______ = 420 / 7

f. _______ = 4,200 / 7

g. 28 / 4 = _______

h. 28,000 / 40 = _______

i. 28,000 / 400 = _______

j. 2,800 / 400 = _______

k. _______ = 24 / 6

l. _______ = 2,400 / 600

m. _______ = 2,400 / 60

n. _______ = 24,000 / 6

Date Time

A Multiples-of-10 Strategy for Division

For Problems 1–4, fill in the multiples-of-10 list. Use the completed list to help answer the question. Then write a number model.

1. The refreshment stand sold 276 cans of soda this weekend. How many 6-packs is that?

   10 [6s] = ______

   20 [6s] = ______

   30 [6s] = ______

   40 [6s] = ______

   50 [6s] = ______

   Answer: ______ 6-packs

   Number model: $276 \div 6 =$ ______

2. We have tables that each seat 4 people. How many tables are needed to seat 178 people?

   10 [4s] = ______

   20 [4s] = ______

   30 [4s] = ______

   40 [4s] = ______

   50 [4s] = ______

   Answer: ______ tables

   Number model: ______ ÷ ______ → ______

3. How many 8s are there in 248?

   10 [8s] = ______

   20 [8s] = ______

   30 [8s] = ______

   40 [8s] = ______

   50 [8s] = ______

   Answer: ______

   Number model: $248 \div 8 =$ ______

4. How many 9s in 431?

   10 [9s] = ______

   20 [9s] = ______

   30 [9s] = ______

   40 [9s] = ______

   50 [9s] = ______

   Answer: ______

   Number model: $\overline{)\quad}$

5. John is 245 days older than Alice. How many weeks older is John?

   Answer: ______ weeks

   Number model: $7\overline{)245}$

6. The preschool held a tricycle parade. Trent counted 72 wheels. How many tricycles is that?

   Answer: ______ tricycles

   Number model: $\overline{)\quad}$

Date _______ Time _______

Multiples of 10, 100, and 1,000

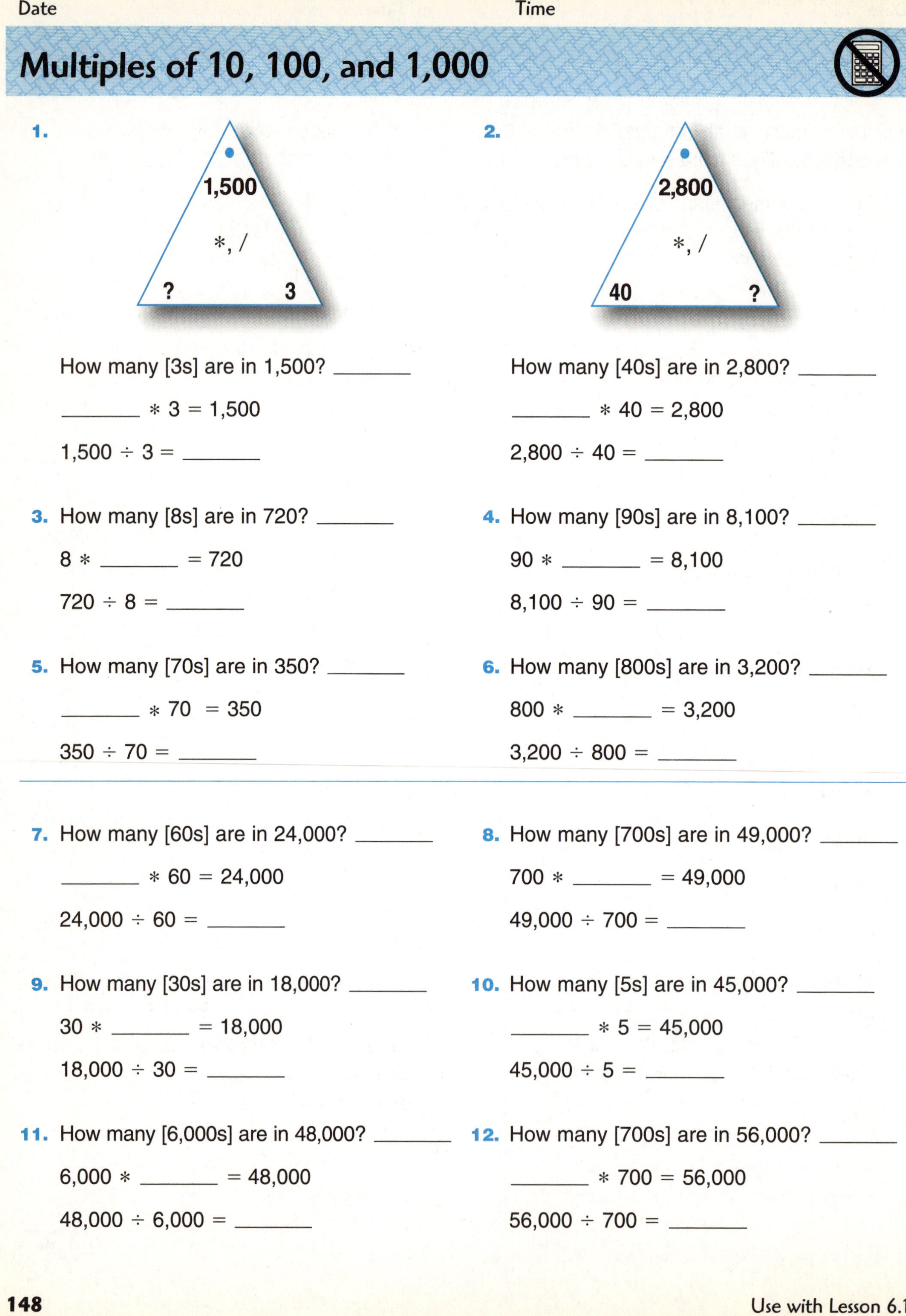

1. How many [3s] are in 1,500? _______

_______ ∗ 3 = 1,500

1,500 ÷ 3 = _______

2. How many [40s] are in 2,800? _______

_______ ∗ 40 = 2,800

2,800 ÷ 40 = _______

3. How many [8s] are in 720? _______

8 ∗ _______ = 720

720 ÷ 8 = _______

4. How many [90s] are in 8,100? _______

90 ∗ _______ = 8,100

8,100 ÷ 90 = _______

5. How many [70s] are in 350? _______

_______ ∗ 70 = 350

350 ÷ 70 = _______

6. How many [800s] are in 3,200? _______

800 ∗ _______ = 3,200

3,200 ÷ 800 = _______

---

7. How many [60s] are in 24,000? _______

_______ ∗ 60 = 24,000

24,000 ÷ 60 = _______

8. How many [700s] are in 49,000? _______

700 ∗ _______ = 49,000

49,000 ÷ 700 = _______

9. How many [30s] are in 18,000? _______

30 ∗ _______ = 18,000

18,000 ÷ 30 = _______

10. How many [5s] are in 45,000? _______

_______ ∗ 5 = 45,000

45,000 ÷ 5 = _______

11. How many [6,000s] are in 48,000? _______

6,000 ∗ _______ = 48,000

48,000 ÷ 6,000 = _______

12. How many [700s] are in 56,000? _______

_______ ∗ 700 = 56,000

56,000 ÷ 700 = _______

Date Time

Math Boxes 6.1

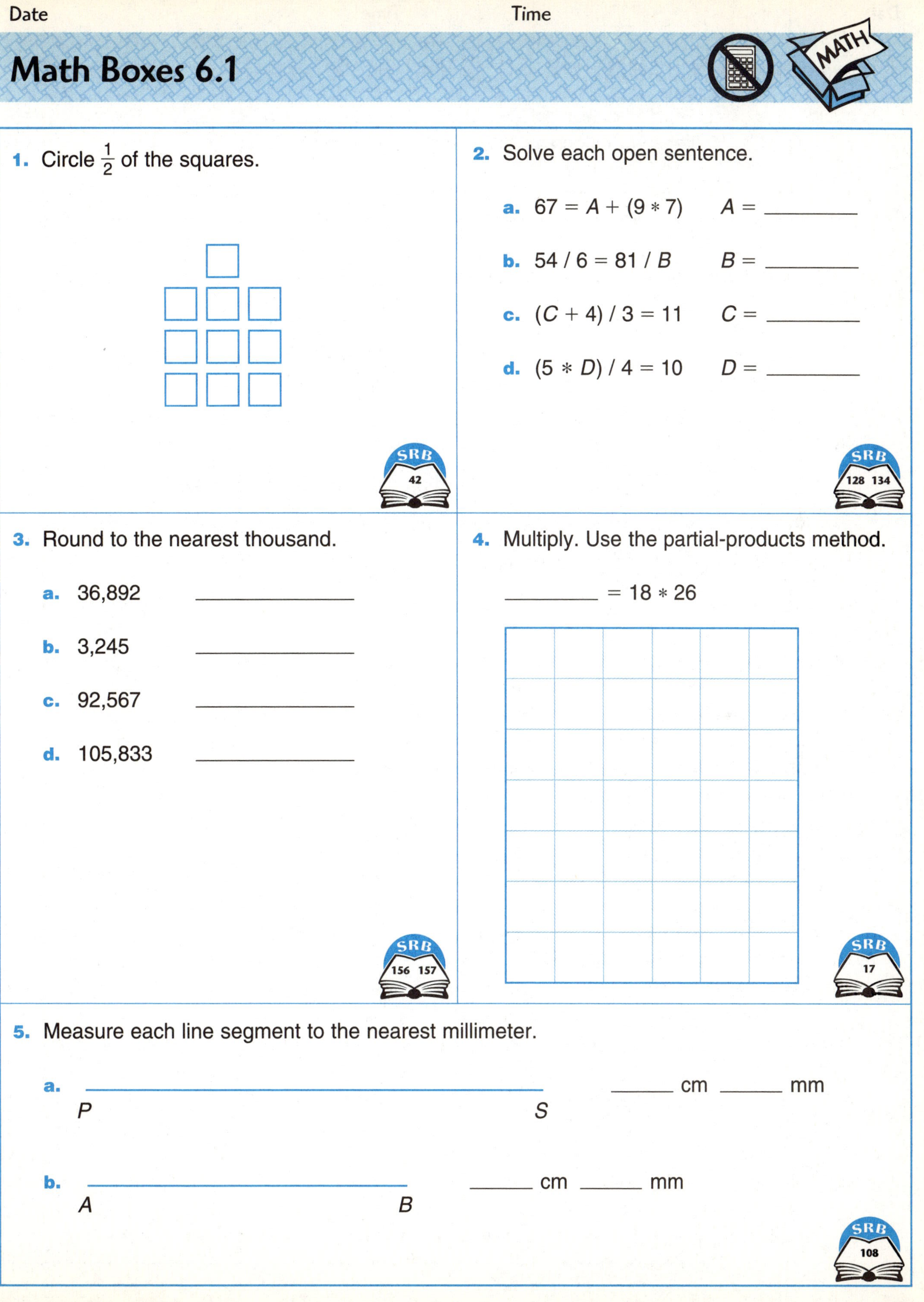

1. Circle $\frac{1}{2}$ of the squares.

SRB 42

2. Solve each open sentence.

a. $67 = A + (9 * 7)$ $A =$ ______

b. $54 / 6 = 81 / B$ $B =$ ______

c. $(C + 4) / 3 = 11$ $C =$ ______

d. $(5 * D) / 4 = 10$ $D =$ ______

SRB 128 134

3. Round to the nearest thousand.

a. 36,892 ______

b. 3,245 ______

c. 92,567 ______

d. 105,833 ______

SRB 156 157

4. Multiply. Use the partial-products method.

______ $= 18 * 26$

SRB 17

5. Measure each line segment to the nearest millimeter.

a. P ______ S ______ cm ______ mm

b. A ______ B ______ cm ______ mm

SRB 108

Date Time

Using the Partial-Quotients Division Algorithm

These notations for division are equivalent:

$12\overline{)246}$ $246 \div 12$ $246 / 12$ $\frac{246}{12}$

Here is a division method you learned in class for dividing 185 by 8.

$8\overline{)185}$		How many 8s are in 185? At least 10.
− 80	10	Use 10 as the first partial quotient. 10 * 8 = 80
105		Subtract. At least 10 [8s] are left.
− 80	10	Use 10 as the second partial quotient. 10 * 8 = 80
25		Subtract. At least 3 [8s] are left.
− 24	3	Use 3 as the third partial quotient. 3 * 8 = 24
1	23	Subtract. Add the partial quotients. 10 + 10 + 3 = 23
↑ **Remainder**	↑ **Quotient**	**Answer: 23 R1**

Divide.

1. $4\overline{)96}$ Answer: ______

2. $147 \div 9$ Answer: ______

3. $253 / 11$ Answer: ______

4. There are 184 plants to be put into pots. Each pot must contain 8 plants. How many pots are needed?

   Answer: ______ pots

   Number model: ______ ÷ ______ = ______

Date Time

Using the Partial-Quotients Division Algorithm (cont.)

5. $14\overline{)462}$ Answer: ____________

6. 331 ÷ 7 Answer: ____________

7. Twelve shirts fit into a box. There are 372 shirts to be put into boxes. How many boxes are needed?

Answer: ________ boxes

Number model: ______ ÷ ______ = ______

8. There are ________ players in the league. (Write a number greater than 100.)

There are ________ players on each team. (Write a number between 3 and 9.)

How many teams can be made?

Answer: ____________ teams

Number model: ______ ÷ ______ → _______

Date Time

Math Boxes 6.2

1. Circle the fractions equivalent to $\frac{1}{2}$.

$\frac{3}{4}$ $\frac{4}{8}$ $\frac{8}{9}$ $\frac{7}{14}$ $\frac{3}{5}$ $\frac{5}{10}$

SRB 49

2. Complete.

a. 670 cm = ________ m

b. 4,800 cm = ________ m

c. 916 cm = ________ m ________ cm

d. 18 m = ________ cm

SRB 109

3. Round 409,381,886 to the nearest

a. hundred. ________________

b. ten-thousand. ________________

c. ten-million. ________________

d. hundred-million. ________________

SRB 156 157

4. Multiply. Use the lattice method.

________ = 86 * 29

2 9
8
6

SRB 18

5. Add or subtract.

a. 0.64 + 1.73 = ________

b. 0.85 + 0.53 = ________

c. ________ = 3.05 – 0.67

d. 12.38 – 1.09 = ________

SRB 32-35

6. Divide.

490 ÷ 8 Answer: ________

SRB 21 22

Date Time

Solving Multiplication and Division Number Stories

Fill in each Multiplication/Division Diagram. Estimate whether the answer will be in the tens, hundreds, or thousands. Then solve the problem and write a number model.

1. The profit from the book sale at Lincoln School was $725. The Math Club and four other clubs will share this amount equally. What will each club's share be?

clubs	dollars per club	dollars in all

Estimate: The answer will be in the TENS HUNDREDS THOUSANDS. (Circle one.)

Answer: ______________________

Number model: ______________________

2. The Box and Bin Store has 475 eight-ounce drinking glasses with "Happy Birthday" on them. These will be packed in boxes of 6, which will sell for $10.98 per box. How many packed boxes will there be?

boxes	glasses per box	glasses in all

Estimate: The answer will be in the TENS HUNDREDS THOUSANDS. (Circle one.)

Answer: ______________________

Number model: ______________________

3. Mr. Day lives 11 miles from his office. Last week he made 16 one-way trips between his home and office. How many miles did he drive in all?

trips	miles per trip	miles in all

Estimate: The answer will be in the TENS HUNDREDS THOUSANDS. (Circle one.)

Answer: ______________________

Number model: ______________________

Date Time

Solving Multiplication and Division Number Stories (cont.)

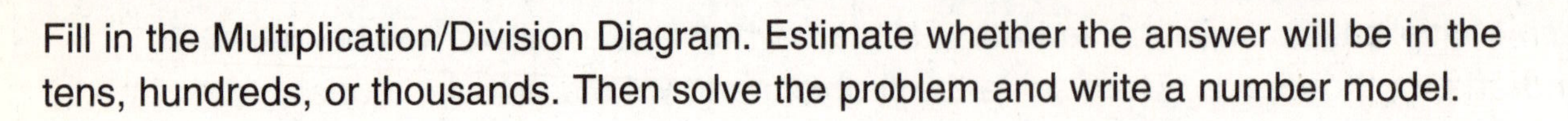

Fill in the Multiplication/Division Diagram. Estimate whether the answer will be in the tens, hundreds, or thousands. Then solve the problem and write a number model.

4. Martina and 3 friends sold 636 boxes of cookies for their club. Each girl sold the same number of boxes. How many boxes of cookies did each girl sell?

girls	boxes per girl	boxes in all

Estimate: The answer will be in the TENS HUNDREDS THOUSANDS. (Circle one.)

Answer: ______________________

Number model: ______________________

Fill in both rows of the Multiplication/Division Diagram. Estimate whether the answer will be in the tens, hundreds, or thousands. Then solve the problem and write a number model.

5. The third and fourth grades received new science books. Eighteen boxes were delivered. Each box contained 12 books. How many books were received?

Estimate: The answer will be in the TENS HUNDREDS THOUSANDS. (Circle one.)

Answer: ______________________

Number model: ______________________

Date Time

Math Boxes 6.3

1. Circle $\frac{5}{6}$ of the squares.

2. Solve each open sentence.

a. $79 = A + (8 * 9)$ $A =$ ______

b. $48 / 6 = 40 / B$ $B =$ ______

c. $(C + 8) / 4 = 7$ $C =$ ______

d. $(6 * D) / 3 = 12$ $D =$ ______

3. Round each to the nearest ten-thousand.

a. 150,983 ______

b. 786,042 ______

c. 12,903,899 ______

d. 79,067,409 ______

4. Multiply. Use the partial-products method.

$54 * 93 =$ ______

5. Measure each line segment to the nearest millimeter.

a. R S

______ cm ______ mm

b. C S

______ cm ______ mm

Date Time

Rewriting and Interpreting Remainders

Example $169 \div 5 \rightarrow 33$ R4

The remainder R4 can be written as $\frac{4}{5}$.

The answer 33 R4 can be written as $33\frac{4}{5}$.

The answer can also be written as a decimal. $33\frac{4}{5} = 33.8$

Write each answer as a mixed number by rewriting the remainder as a fraction.

1. $2\overline{)27}$ → 13 R1 ________
2. $10\overline{)883}$ → 88 R3 ________
3. $16\overline{)252}$ → 15 R12 ________
4. $100\overline{)770}$ → 7 R70 ________

Divide, rewrite the remainder as fraction or decimal, and keep the fraction or decimal as part of the final answer.

5. A board 93 inches long is cut into 12 pieces of equal length. How long is each piece?

   ________ inches

6. It costs $50 to buy 8 adult tickets to the school play. What is the cost per ticket?

   $ ________

Solve. Be prepared to tell what you did when there was a remainder—ignored it, made it part of the answer, or rounded up the answer.

7. The school library is buying boxes to store 128 videotapes. Each box holds 6 tapes. How many boxes are needed to store all of the tapes?

   ________ boxes

8. Jackson is having a party. Balloons cost $6 per bunch. How many bunches can he buy with $75?

   ________ bunches

9. There are 196 note cards. They are shared equally by 14 students. How many note cards does each student get?

   ________ note cards

10. Eight $\frac{1}{2}$-hour guitar lessons cost $60. What is the cost per lesson?

    $ ________

Date Time

Writing Multiplication and Division Number Stories

The following number models can be used to help solve number stories. For each, write a number story. Then ask your partner to solve it and write the solution, including any units. At the same time, solve your partner's story.

1. $4 * 36 = ?$

Number story: ______________________________

____

____

____

Solution: ______________________________ (unit)

2. $8\overline{)360} = ?$

Number story: ______________________________

____

____

____

Solution: ______________________________ (unit)

3. $5,280 / 6 = ?$

Number story: ______________________________

____

____

____

Solution: ______________________________ (unit)

Date Time

Math Boxes 6.4

1. Circle the fractions equivalent to $\frac{1}{2}$.

$\frac{8}{16}$ $\frac{5}{6}$ $\frac{6}{12}$ $\frac{2}{3}$ $\frac{12}{24}$ $\frac{8}{15}$

2. Complete.

a. 320 cm = ________ m

b. 5,600 cm = ________ m

c. 412 cm = ________ m ________ cm

d. 12 m = ________ cm

3. Round 5,906,245 to the nearest

a. million. ____________

b. ten-thousand. ____________

c. thousand. ____________

d. hundred. ____________

4. Multiply. Use the lattice method.

________ = 58 * 52

5. Add or subtract.

a. 2.01 + 5.01 = ________

b. 0.37 + 0.26 = ________

c. ________ = 7.80 – 3.65

d. 6.79 – 6.55 = ________

6. Divide.

872 ÷ 5 Answer: ________

Date Time

A Map of the Island of Ireland

Bantry	B–1	Dublin	F–4	Lahinch	B–4	Omagh	E–7
Belfast	F–7	Dundalk	F–6	Larne	F–7	Tralee	B–2
Carlow	E–3	Galway	C–4	Limerick	C–3	Tuam	C–5
Castlebar	B–6	Gort	C–4	Mullingar	E–5	Westport	B–5
Derry	E–8	Kilkee	B–3	Navan	E–5	Wicklow	F–4

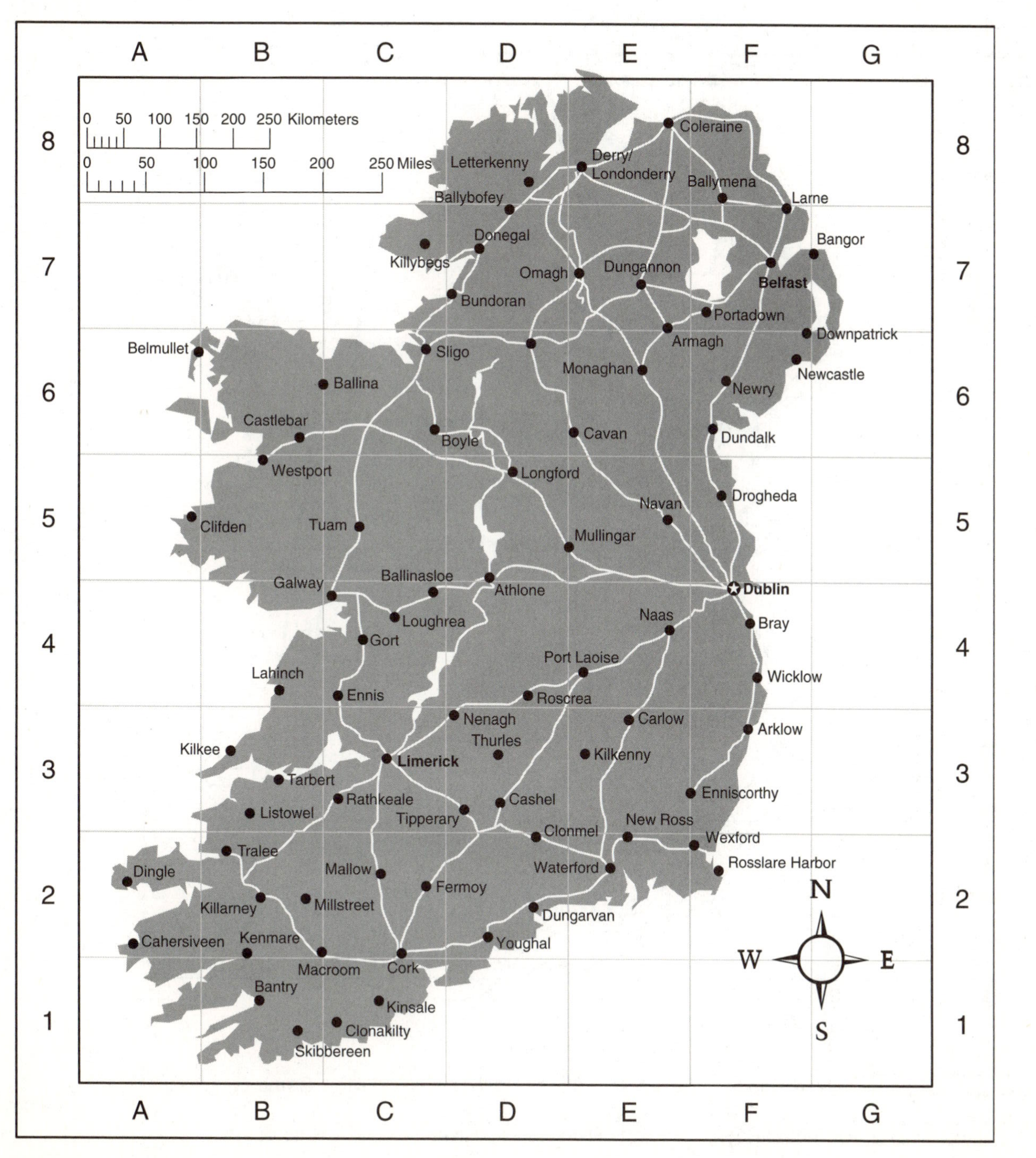

Date Time

A Campground Map

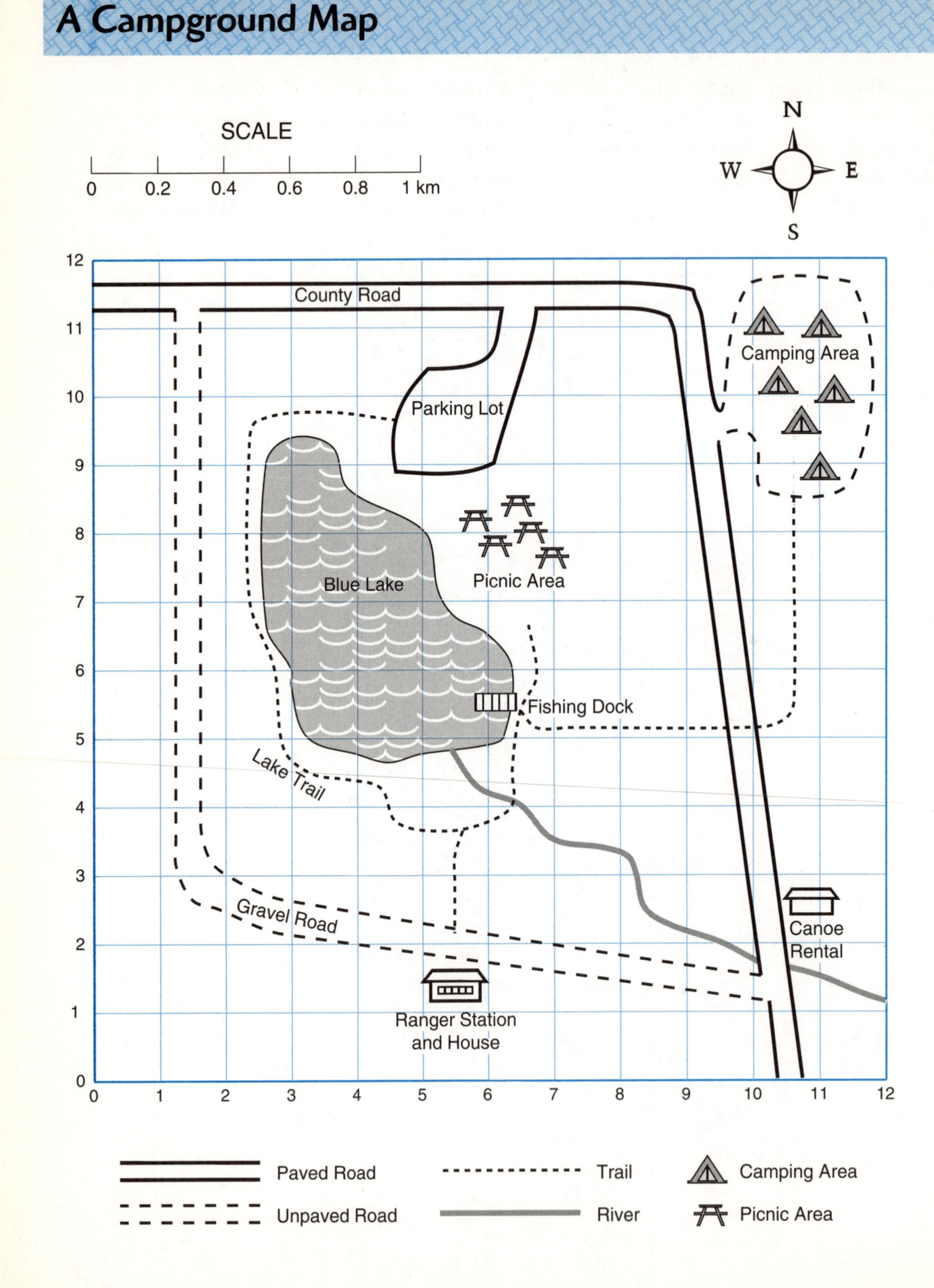

Use with Lesson 6.5.

Date Time

Finding Distances on a Map

Use the campground map on journal page 160 to complete the following:

1. Suppose you hiked along the lake trail from the fishing dock to the parking lot. Estimate the distance you hiked. About ________ km

2. The ranger made her hourly check. She started at the ranger station. She drove northwest and then north on Gravel Road to County Road. She turned east onto County Road and drove past the parking lot and the camping area. After she passed the canoe rental, she turned right onto Gravel Road and drove back to the ranger station. About what distance did she drive? About ________ km

3. Estimate the distance around Blue Lake. About ________ km

4. You are planning to hike from the camping area to the parking lot. You will stay on the roads or trails. You want to hike at least 5 kilometers.

   a. Plan your route. Then draw it on the map with a coloring pencil or crayon.

   b. Estimate the distance. About ________ km

5. Use the ordered number pairs to locate each item on the map. Mark a dot at the location. Next to the dot, write the letter given for the item.

Campground Features Chart

	Location	Letter
parked car	(5,9)	C
boat	$(3\frac{1}{2},8)$	B
swing set	(8,11)	S
hikers	(10.5,6.5)	H
farmhouse	$(\frac{1}{2},7)$	F

Date Time

Rewriting Remainders

Write each answer as a mixed number by rewriting the remainder as a fraction.

1. $2\overline{)79}$ = 39 R1 ________
2. $10\overline{)446}$ = 44 R6 ________
3. $8\overline{)394}$ = 49 R2 ________
4. $16\overline{)860}$ = 53 R12 ________

Divide. Write each answer as a mixed number by writing the remainder as a fraction.

5. $427 \div 2$ **Answer:** ________

6. 390 / 9 **Answer:** ________

7. $12\overline{)675}$ **Answer:** ________

Date Time

Math Boxes 6.5

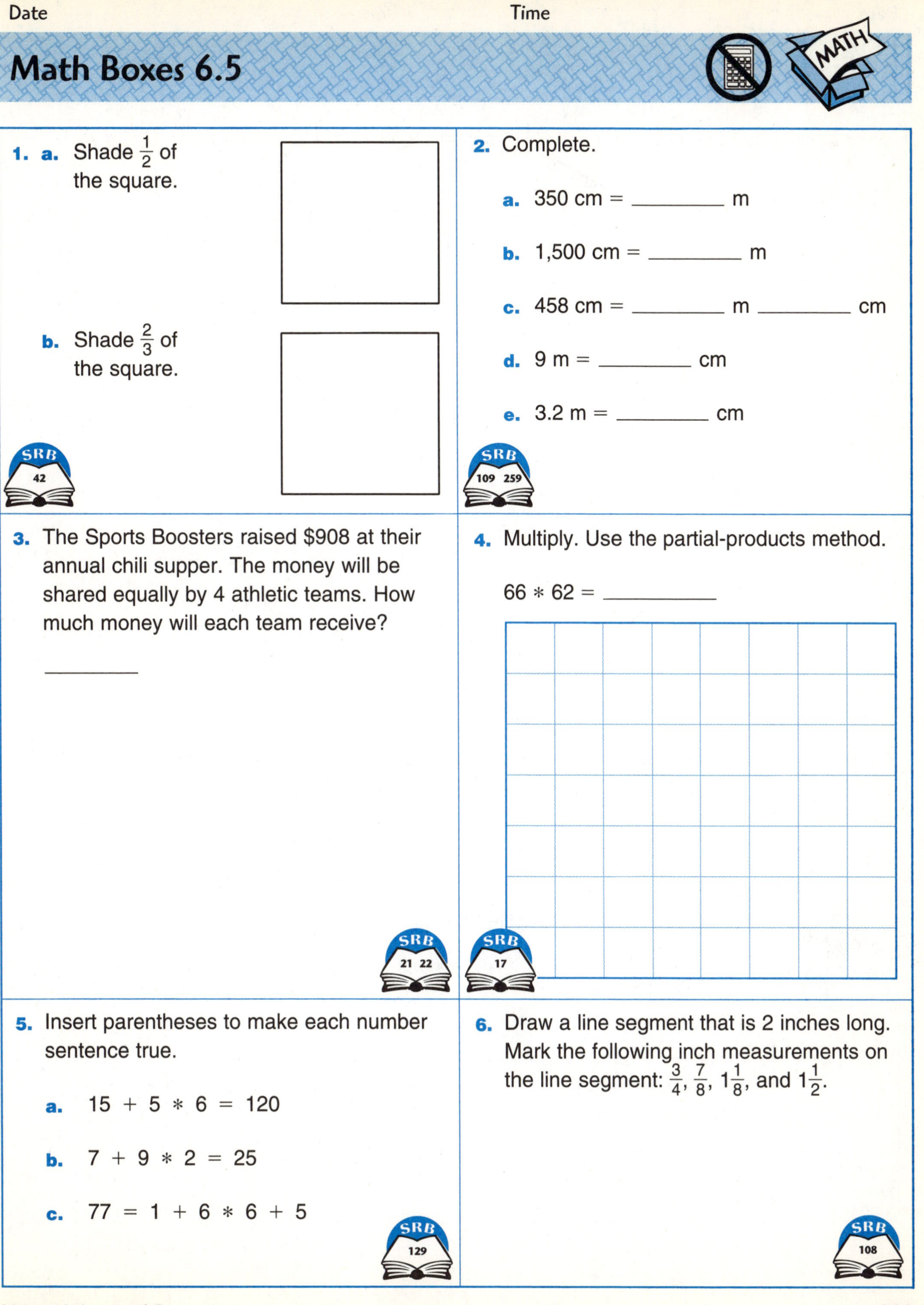

1. a. Shade $\frac{1}{2}$ of the square.

   b. Shade $\frac{2}{3}$ of the square.

SRB 42

2. Complete.

   a. 350 cm = ________ m

   b. 1,500 cm = ________ m

   c. 458 cm = ________ m ________ cm

   d. 9 m = ________ cm

   e. 3.2 m = ________ cm

SRB 109 259

3. The Sports Boosters raised $908 at their annual chili supper. The money will be shared equally by 4 athletic teams. How much money will each team receive?

   ____

SRB 21 22

4. Multiply. Use the partial-products method.

   66 * 62 = ________

SRB 17

5. Insert parentheses to make each number sentence true.

   a. 15 + 5 * 6 = 120

   b. 7 + 9 * 2 = 25

   c. 77 = 1 + 6 * 6 + 5

SRB 129

6. Draw a line segment that is 2 inches long. Mark the following inch measurements on the line segment: $\frac{3}{4}$, $\frac{7}{8}$, $1\frac{1}{8}$, and $1\frac{1}{2}$.

SRB 108

Date Time

Making an Angle Measurer

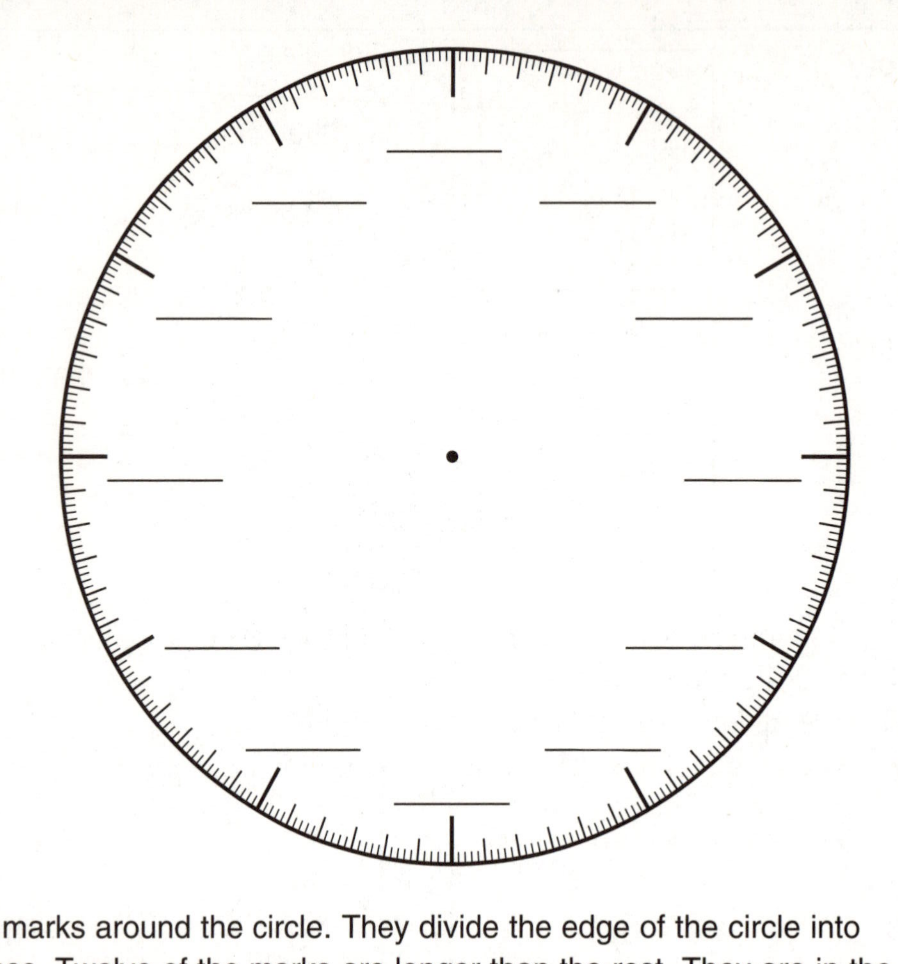

There are 360 marks around the circle. They divide the edge of the circle into 360 small spaces. Twelve of the marks are longer than the rest. They are in the same positions as the 12 numbers around a clock face. Your teacher will tell you how to label the 12 large marks on the circle.

The Babylonian Number System

The Babylonians lived about 3,000 years ago in what is now the country of Iraq. Babylon was on the Euphrates River. It was not far from Baghdad, which is the current capital of Iraq.

The Babylonians used a numbering system based on the number 60. Today, Babylonian ideas are used to measure time. The hour is divided into 60 minutes, and the minute into 60 seconds.

The number 60 is also used in measuring angles. A full turn measures 360°, and 360° is a multiple of 60° ($6 * 60° = 360°$).

Date ______ Time ______

Clock Angles

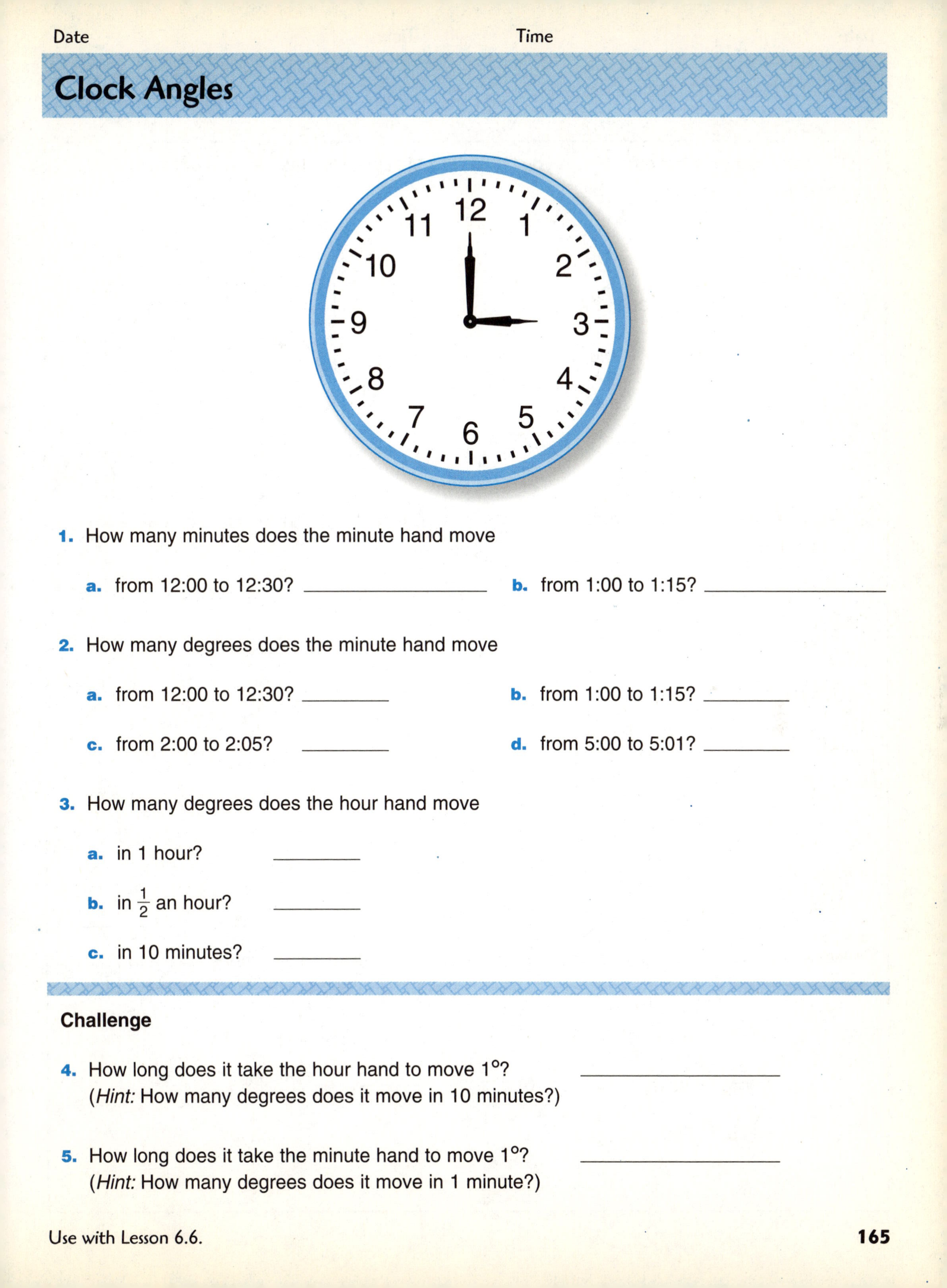

1. How many minutes does the minute hand move

   a. from 12:00 to 12:30? ________ b. from 1:00 to 1:15? ________

2. How many degrees does the minute hand move

   a. from 12:00 to 12:30? ________ b. from 1:00 to 1:15? ________

   c. from 2:00 to 2:05? ________ d. from 5:00 to 5:01? ________

3. How many degrees does the hour hand move

   a. in 1 hour? ________

   b. in $\frac{1}{2}$ an hour? ________

   c. in 10 minutes? ________

Challenge

4. How long does it take the hour hand to move 1°? ________
   (*Hint:* How many degrees does it move in 10 minutes?)

5. How long does it take the minute hand to move 1°? ________
   (*Hint:* How many degrees does it move in 1 minute?)

Date Time

Math Boxes 6.6

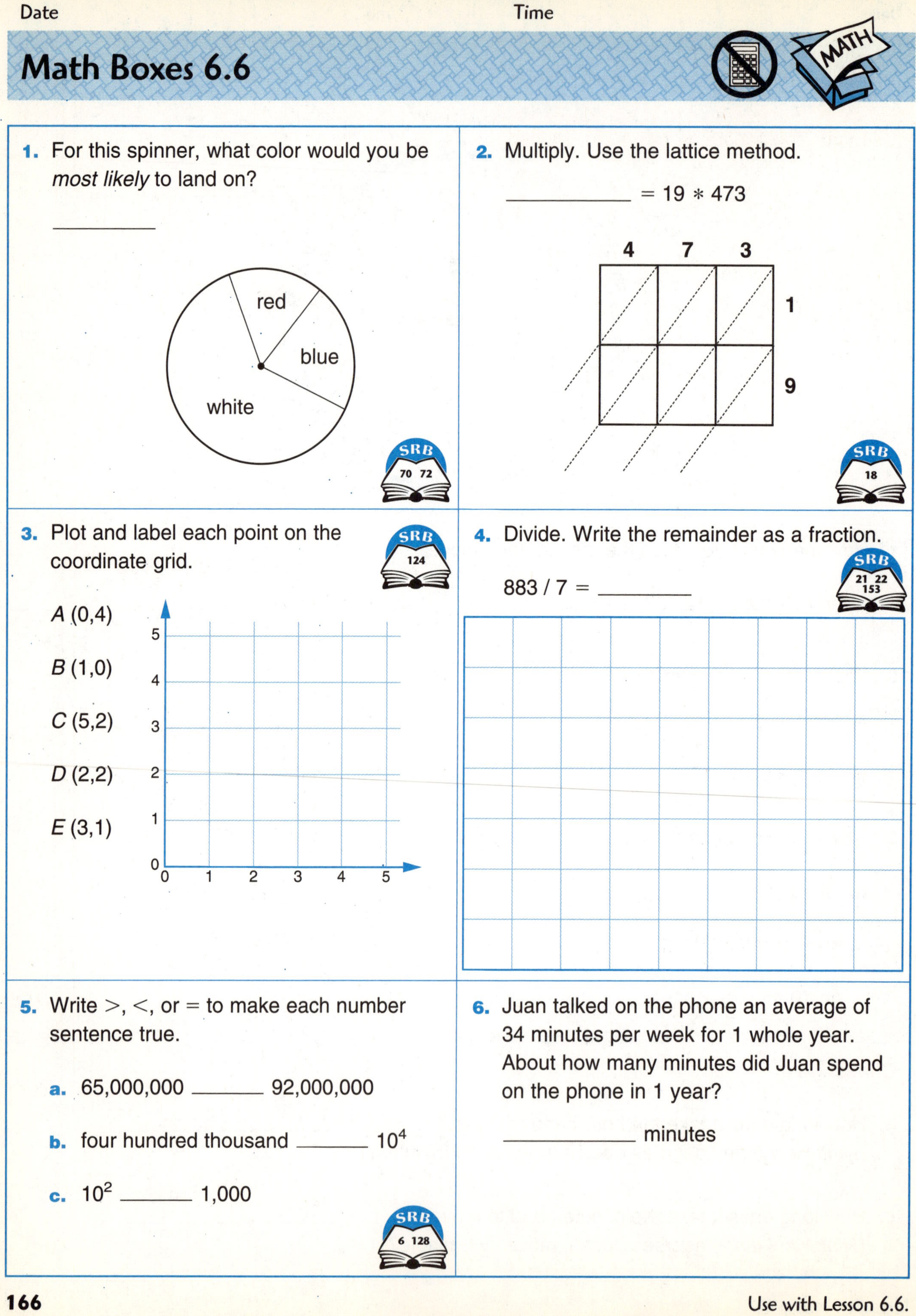

1. For this spinner, what color would you be *most likely* to land on?

____

red
blue
white

SRB 70 72

2. Multiply. Use the lattice method.

_______ $= 19 * 473$

4 7 3
1
9

SRB 18

3. Plot and label each point on the coordinate grid.

A (0,4)

B (1,0)

C (5,2)

D (2,2)

E (3,1)

0 1 2 3 4 5

SRB 124

4. Divide. Write the remainder as a fraction.

883 / 7 = _______

SRB 21 22 153

5. Write >, <, or = to make each number sentence true.

a. 65,000,000 _______ 92,000,000

b. four hundred thousand _______ 10^4

c. 10^2 _______ 1,000

SRB 6 128

6. Juan talked on the phone an average of 34 minutes per week for 1 whole year. About how many minutes did Juan spend on the phone in 1 year?

_______ minutes

Date Time

Measuring Angles with a 360° Angle Measurer

Use your angle measurer to measure each angle.

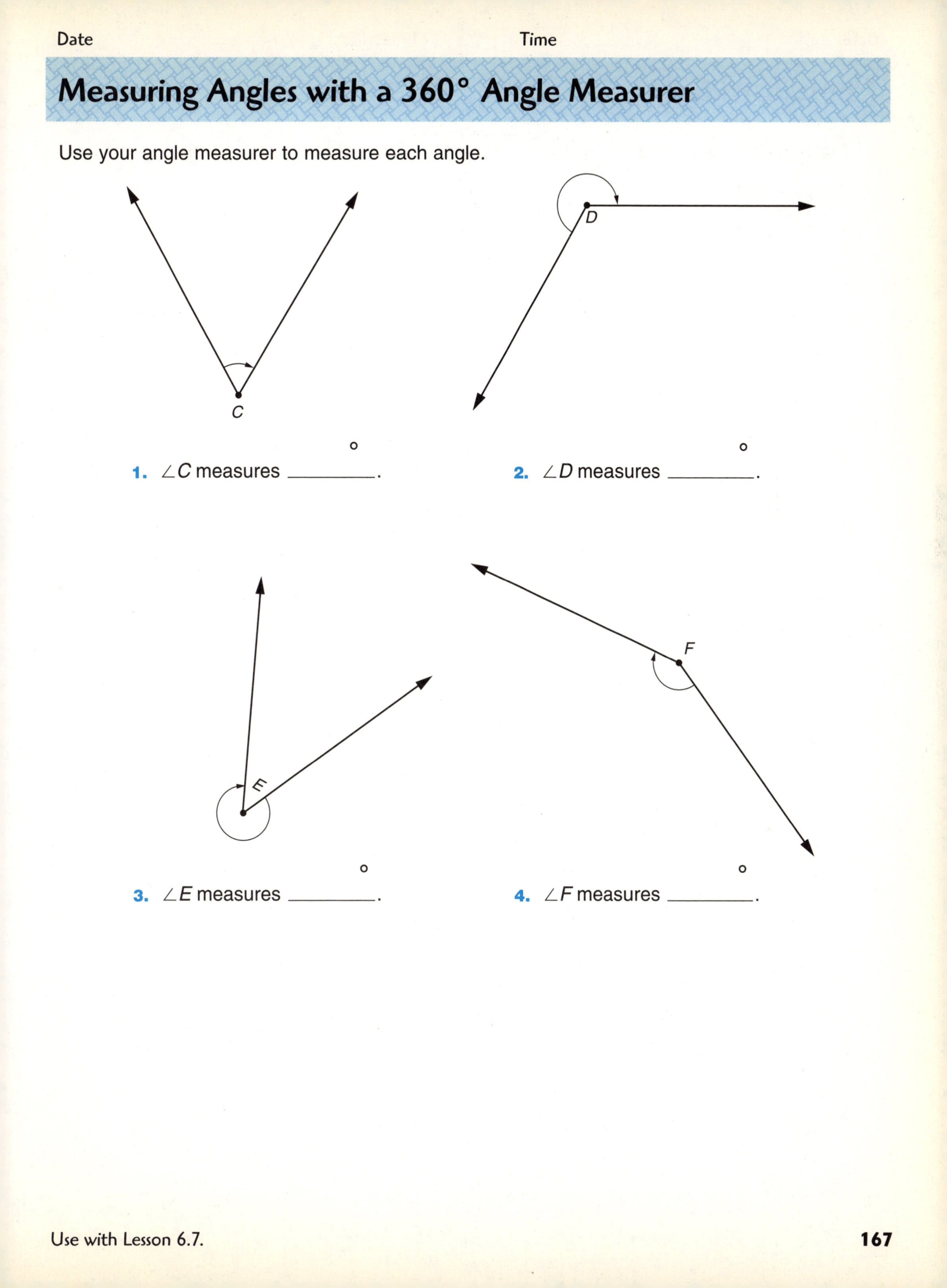

1. $\angle C$ measures ________°.

2. $\angle D$ measures ________°.

3. $\angle E$ measures ________°.

4. $\angle F$ measures ________°.

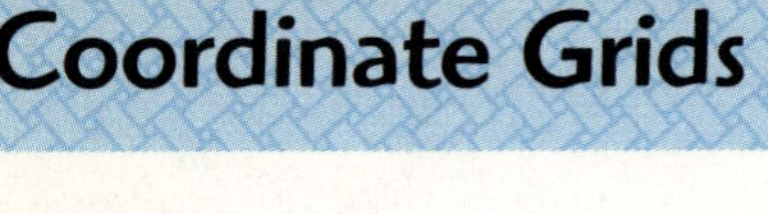

Coordinate Grids

1. Plot and label each point on the coordinate grid.

A	(1,7)
B	(6,6)
C	(10,1)
D	(4,3)
E	(8,6)
F	(2,9)
G	(9,1)
H	(10,4)

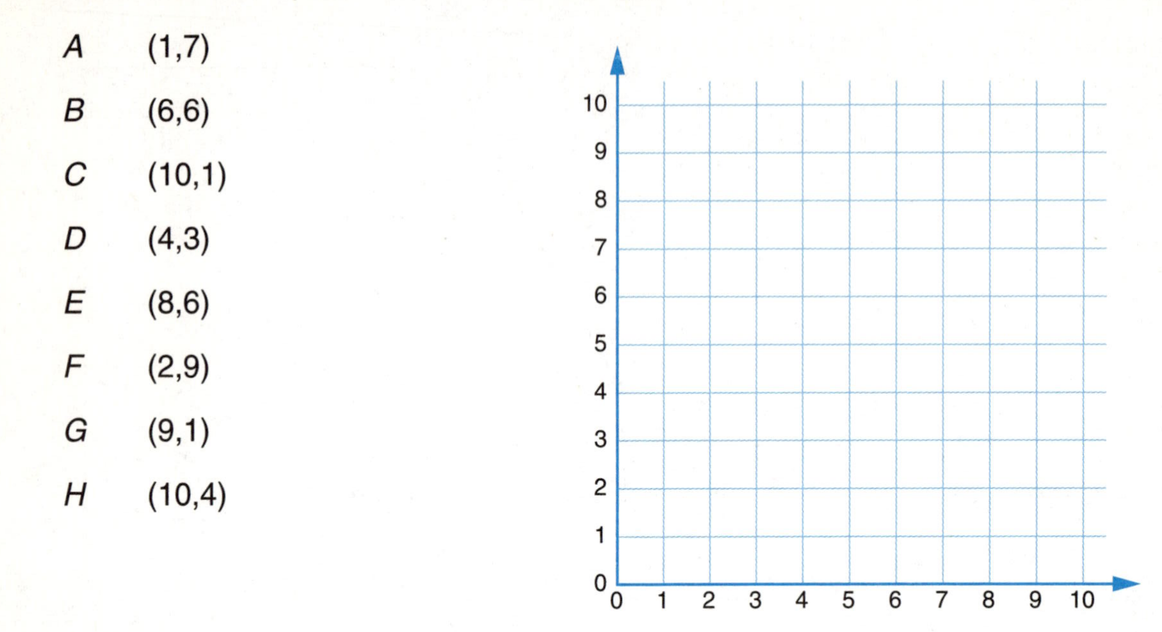

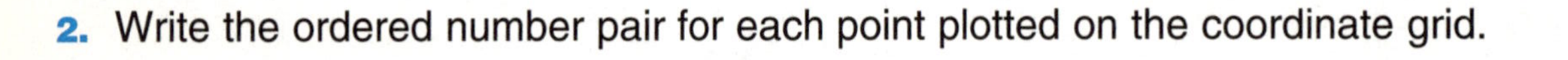

2. Write the ordered number pair for each point plotted on the coordinate grid.

I (_____ , _____)

J (_____ , _____)

K (_____ , _____)

L (_____ , _____)

M (_____ , _____)

N (_____ , _____)

O (_____ , _____)

P (_____ , _____)

Q (_____ , _____)

R (_____ , _____)

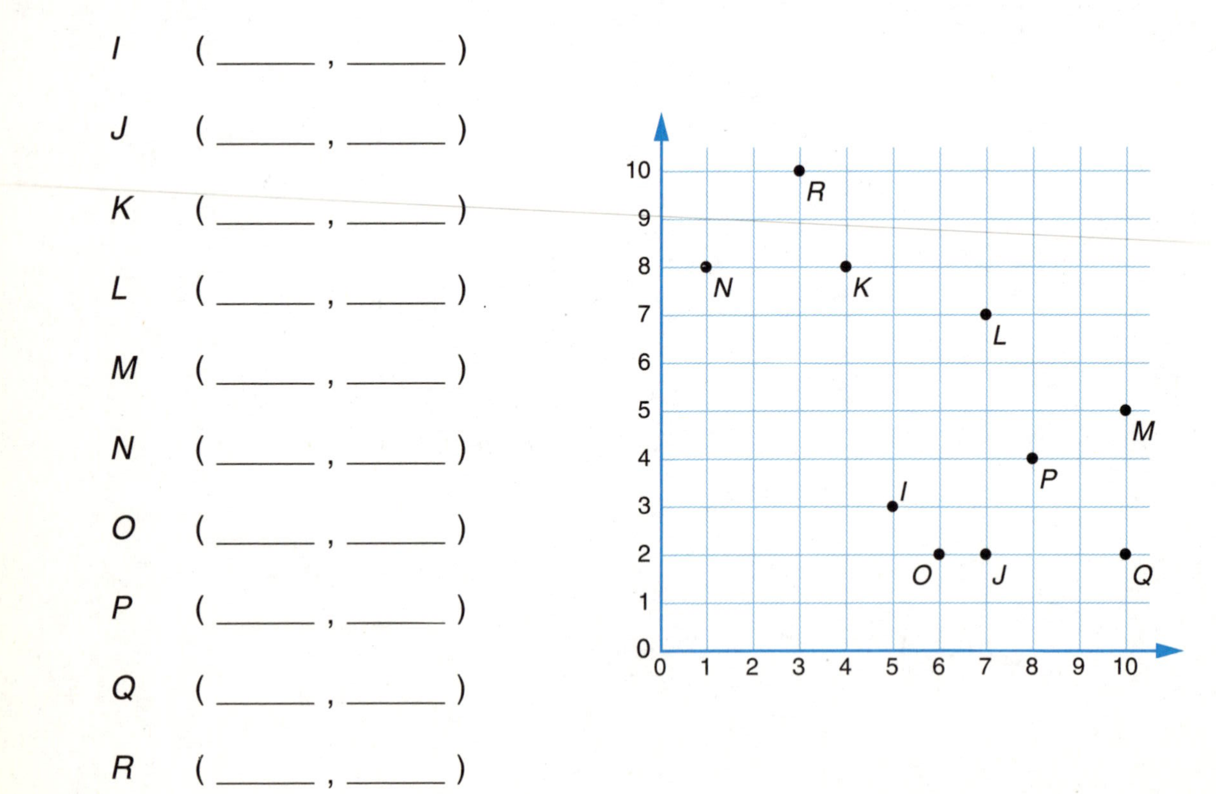

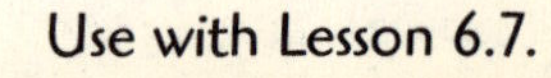

Date Time

Math Boxes 6.7

1. a. Shade $\frac{1}{3}$ of the rectangle.

   b. Shade $\frac{3}{4}$ of the rectangle.

2. Complete.

   a. 750 cm = ________ m

   b. 3,500 cm = ________ m

   c. 797 cm = ________ m ________ cm

   d. 8 m = ________ cm

   e. 9.7 m = ________ cm

3. 27 fourth grade students collected newspapers for one week. If they collected a total of 2,078 newspapers by the end of the week, on average about how many papers did each student collect?

   ________ newspapers

4. Multiply. Use the partial-products method.

   $67 * 34 =$ ____________

5. Insert parentheses to make each number sentence true.

   a. $66 - 16 * 4 = 200$

   b. $49 = 4 + 3 * 42 / 6$

   c. $12 = 15 - 2 + 1$

6. Draw a line segment that is $1\frac{3}{4}$ inches long. Mark the following inch measurements on the line segment: $\frac{1}{4}$, $\frac{5}{8}$, $1\frac{3}{8}$, and $1\frac{1}{2}$.

Date Time

Drawing and Measuring Angles

Math Message

Use a straightedge to draw the following angles. Do **not** use an angle measurer.

$\angle A$: any angle less than 90°	$\angle B$: any angle more than 90° and less than 180°	$\angle C$: any angle more than 180°
$\angle A$ is called an **acute angle.**	$\angle B$ is called an **obtuse angle.**	$\angle C$ is called a **reflex angle.**

Measuring Angles with a Protractor

Measure each angle as accurately as you can.

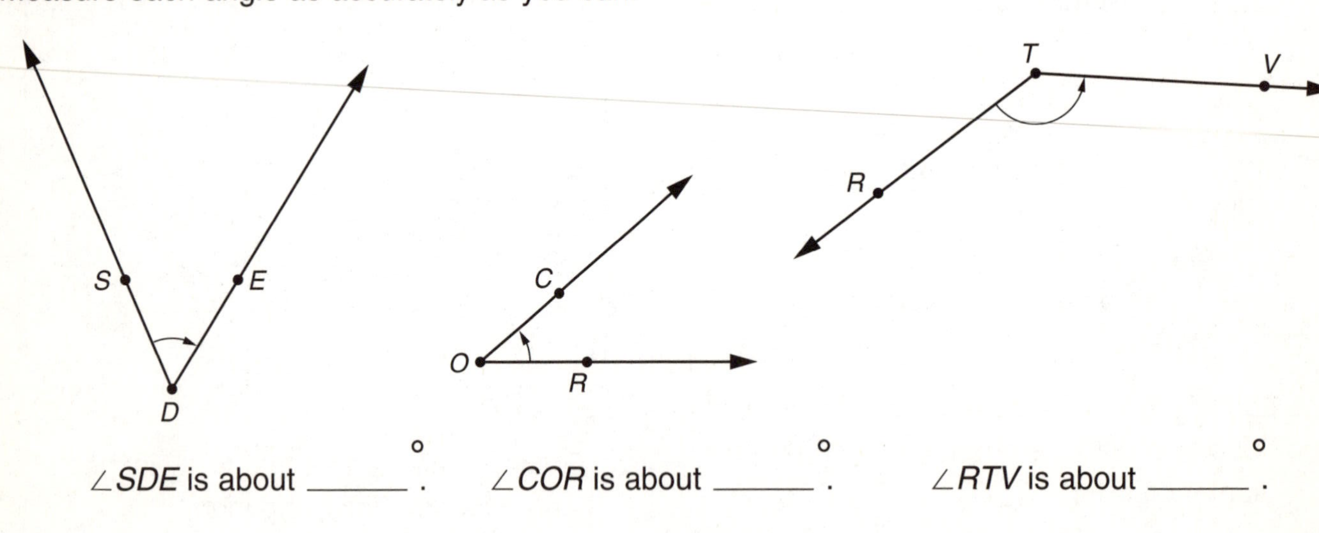

$\angle SDE$ is about ______ °. $\angle COR$ is about ______ °. $\angle RTV$ is about ______ °.

Date Time

Exploring Triangle Measures

Work with a partner. You will each need a sheet of paper, a straightedge, and a protractor.

1. Each partner draws a large triangle on a separate sheet of paper. The two triangles should not look the same.

2. Label the vertices of one triangle *A, B,* and *C.* Label the vertices of the other triangle *D, E,* and *F.*

3. Use your protractors to measure each angle as accurately as you can.

4. Record the degree measures in the tables below.

5. Find the sum of the degree measures of triangle *ABC* and triangle *DEF.*

Angle	Degree Measure
$\angle A$	About ______ °
$\angle B$	About ______ °
$\angle C$	About ______ °
Sum	About ______ °

Angle	Degree Measure
$\angle D$	About ______ °
$\angle E$	About ______ °
$\angle F$	About ______ °
Sum	About ______ °

6. Write a true statement about the three angles of a triangle.

____

____

____

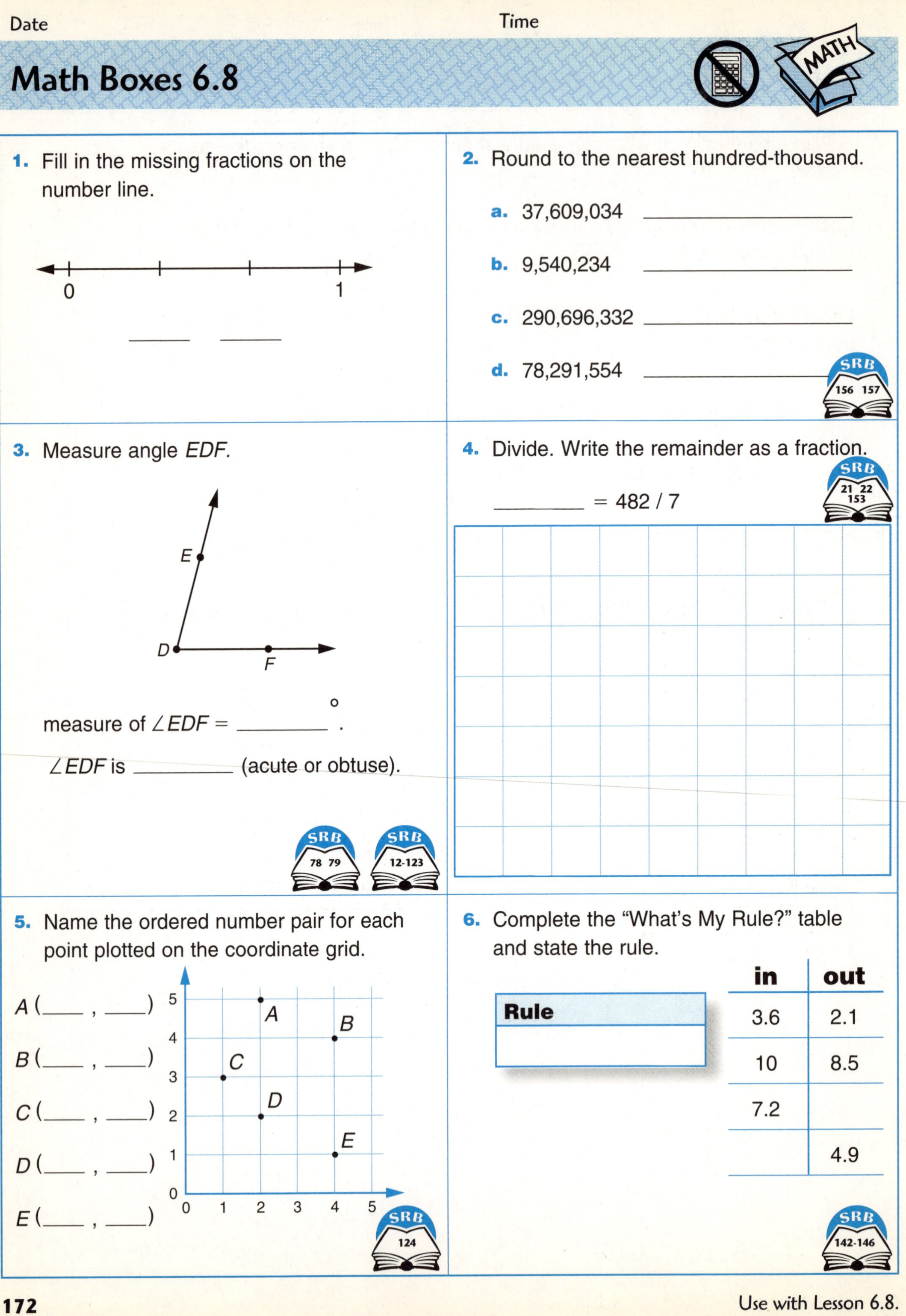
Date
Time
Math Boxes 6.8
MATH
1. Fill in the missing fractions on the number line.
0
1
2. Round to the nearest hundred-thousand.
a. 37,609,034
b. 9,540,234
c. 290,696,332
d. 78,291,554
SRB 156 157
3. Measure angle EDF.
E
D
F
measure of ∠EDF = ____°.
∠EDF is ____ (acute or obtuse).
SRB 78 79
SRB 12-123
4. Divide. Write the remainder as a fraction.
____ = 482 / 7
SRB 21 22 153
5. Name the ordered number pair for each point plotted on the coordinate grid.
A (___ , ___)
B (___ , ___)
C (___ , ___)
D (___ , ___)
E (___ , ___)
0
1
2
3
4
5
A
B
C
D
E
SRB 124
6. Complete the "What's My Rule?" table and state the rule.
Rule
in
out
3.6
2.1
10
8.5
7.2
4.9
SRB 142-146

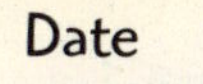

Date Time

Drawing Angles

1. Draw a 15° angle, using ray *AB* as one of its sides.

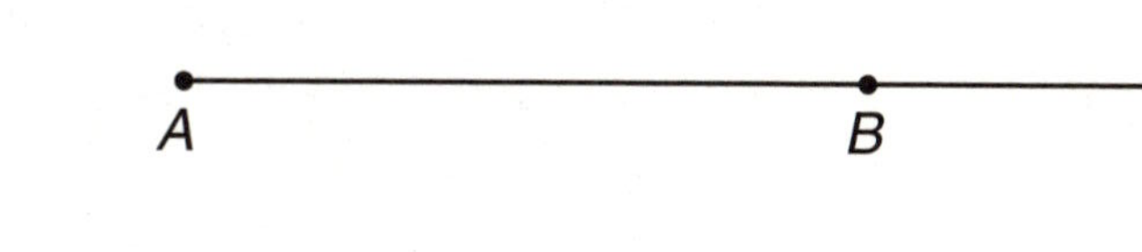

2. Draw a 150° angle, using ray *CD* as one of its sides.

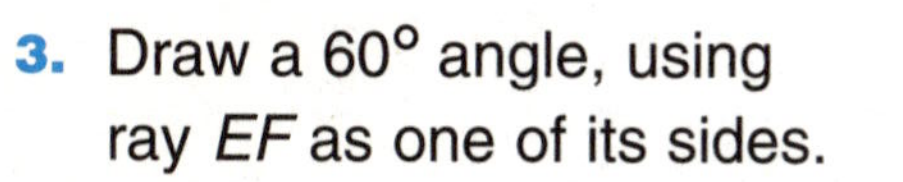

3. Draw a 60° angle, using ray *EF* as one of its sides.

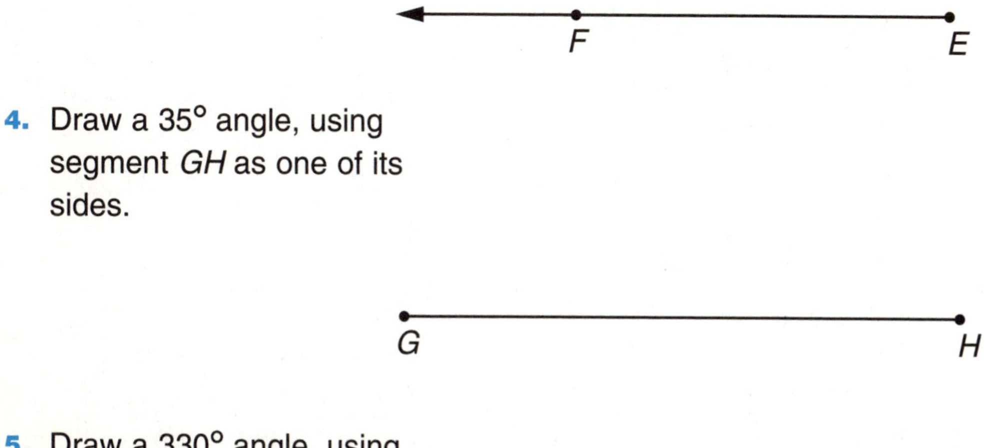

4. Draw a 35° angle, using segment *GH* as one of its sides.

5. Draw a 330° angle, using ray *IJ* as one of its sides.

I J

Date _______ Time _______

Math Boxes 6.9

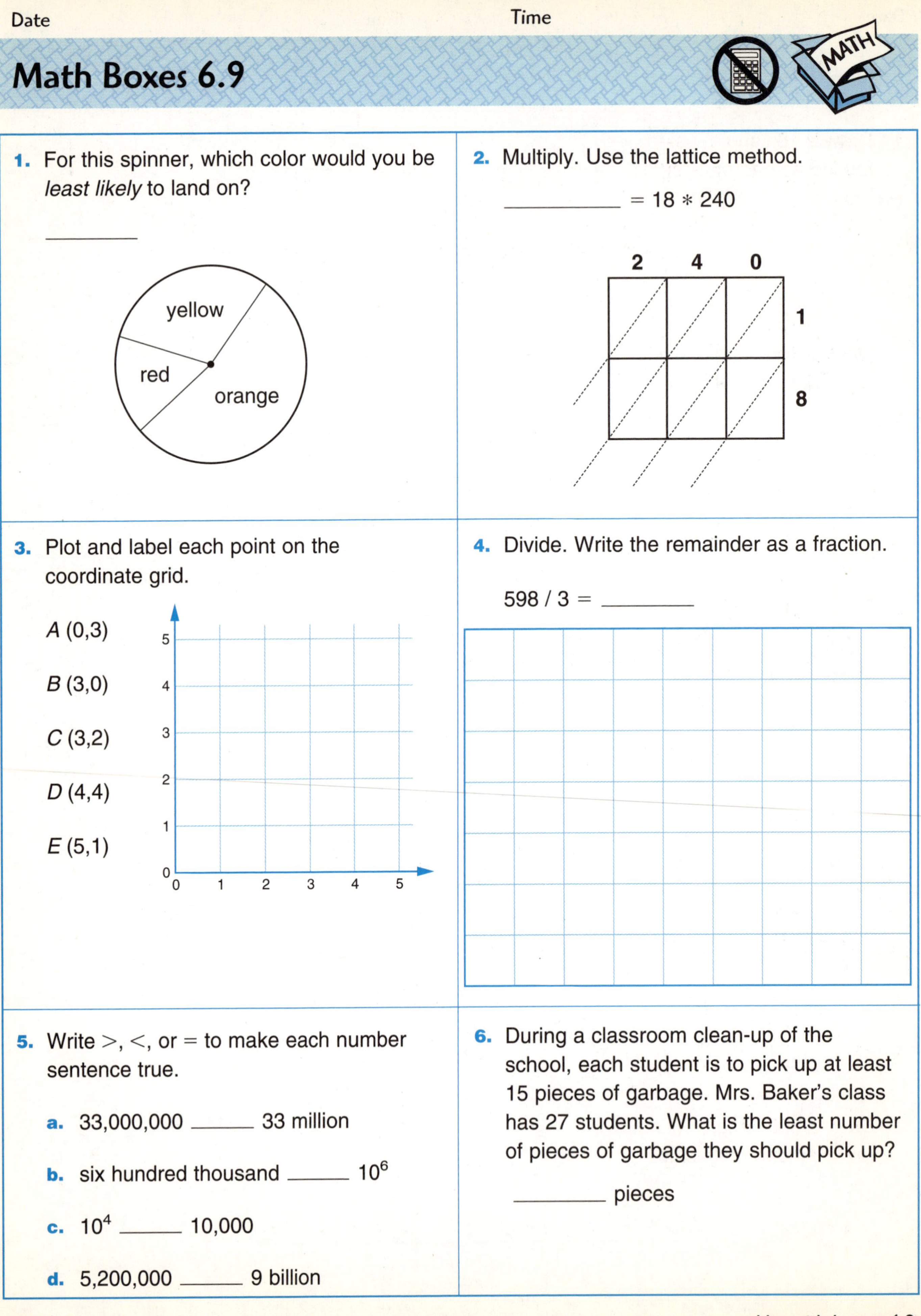

1. For this spinner, which color would you be *least likely* to land on?

____

2. Multiply. Use the lattice method.

_______ $= 18 * 240$

3. Plot and label each point on the coordinate grid.

A (0,3)

B (3,0)

C (3,2)

D (4,4)

E (5,1)

4. Divide. Write the remainder as a fraction.

598 / 3 = _______

5. Write >, <, or = to make each number sentence true.

a. 33,000,000 _______ 33 million

b. six hundred thousand _______ 10^6

c. 10^4 _______ 10,000

d. 5,200,000 _______ 9 billion

6. During a classroom clean-up of the school, each student is to pick up at least 15 pieces of garbage. Mrs. Baker's class has 27 students. What is the least number of pieces of garbage they should pick up?

_______ pieces

Date Time

Locating Places on Regional Maps

1. Find each of the following cities on the maps in the World Tour section of your *Student Reference Book.* Record the continent in which it is located.

a. Lisbon, Portugal (on Region 2 map) ____________________

b. Mumbai (Bombay), India (on Region 4 map) ____________________

c. La Paz, Bolivia (on Region 3 map) ____________________

d. Pretoria, South Africa ____________________

e. Vancouver, Canada ____________________

2. Use the maps of continents to find the approximate latitude and longitude of each city.

a. Lisbon, Portugal latitude _____° North; longitude _____° West

b. Mumbai, India latitude _____° _______; longitude _____° _______

c. La Paz, Bolivia latitude _____° _______; longitude _____° _______

d. Pretoria, S. Africa latitude _____° _______; longitude _____° _______

e. Vancouver, Canada latitude _____° _______; longitude _____° _______

3. Each degree of latitude (traveling north or south from the equator) is equal to about 70 miles. About how many miles from the equator is each city listed below?

a. Lisbon, Portugal About ________ miles

b. Mumbai, India About ________ miles

c. La Paz, Bolivia About ________ miles

d. Pretoria, South Africa About ________ miles

e. Vancouver, Canada About ________ miles

Date Time

Math Boxes 6.10

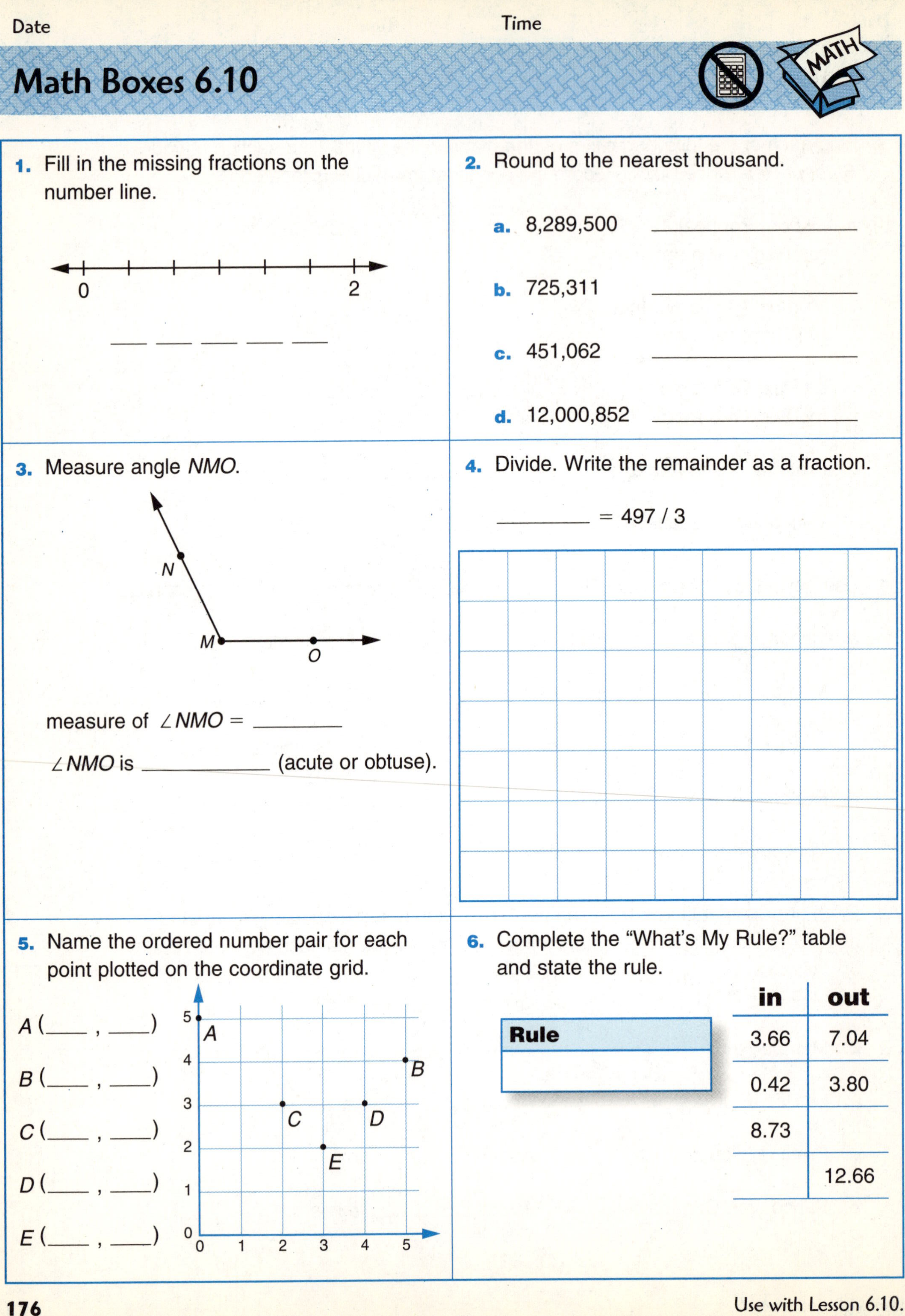

1. Fill in the missing fractions on the number line.

0 ___ ___ ___ ___ ___ 2

2. Round to the nearest thousand.

   a. 8,289,500 ________

   b. 725,311 ________

   c. 451,062 ________

   d. 12,000,852 ________

3. Measure angle *NMO*.

measure of $\angle NMO$ = ______

$\angle NMO$ is ________ (acute or obtuse).

4. Divide. Write the remainder as a fraction.

______ = 497 / 3

5. Name the ordered number pair for each point plotted on the coordinate grid.

A (___ , ___)

B (___ , ___)

C (___ , ___)

D (___ , ___)

E (___ , ___)

6. Complete the "What's My Rule?" table and state the rule.

Rule

in	out
3.66	7.04
0.42	3.80
8.73	
	12.66

Time to Reflect

1. Do you think division is easy or hard? Explain your answer.

2. Write 5 sentences that each contain the word *angle.* At least 2 of the sentences must be about mathematics.

a.

b.

c.

d.

e.

3. After working in this unit, do you find it easier to find places on a globe? Why or why not?

Date Time

Math Boxes 6.11

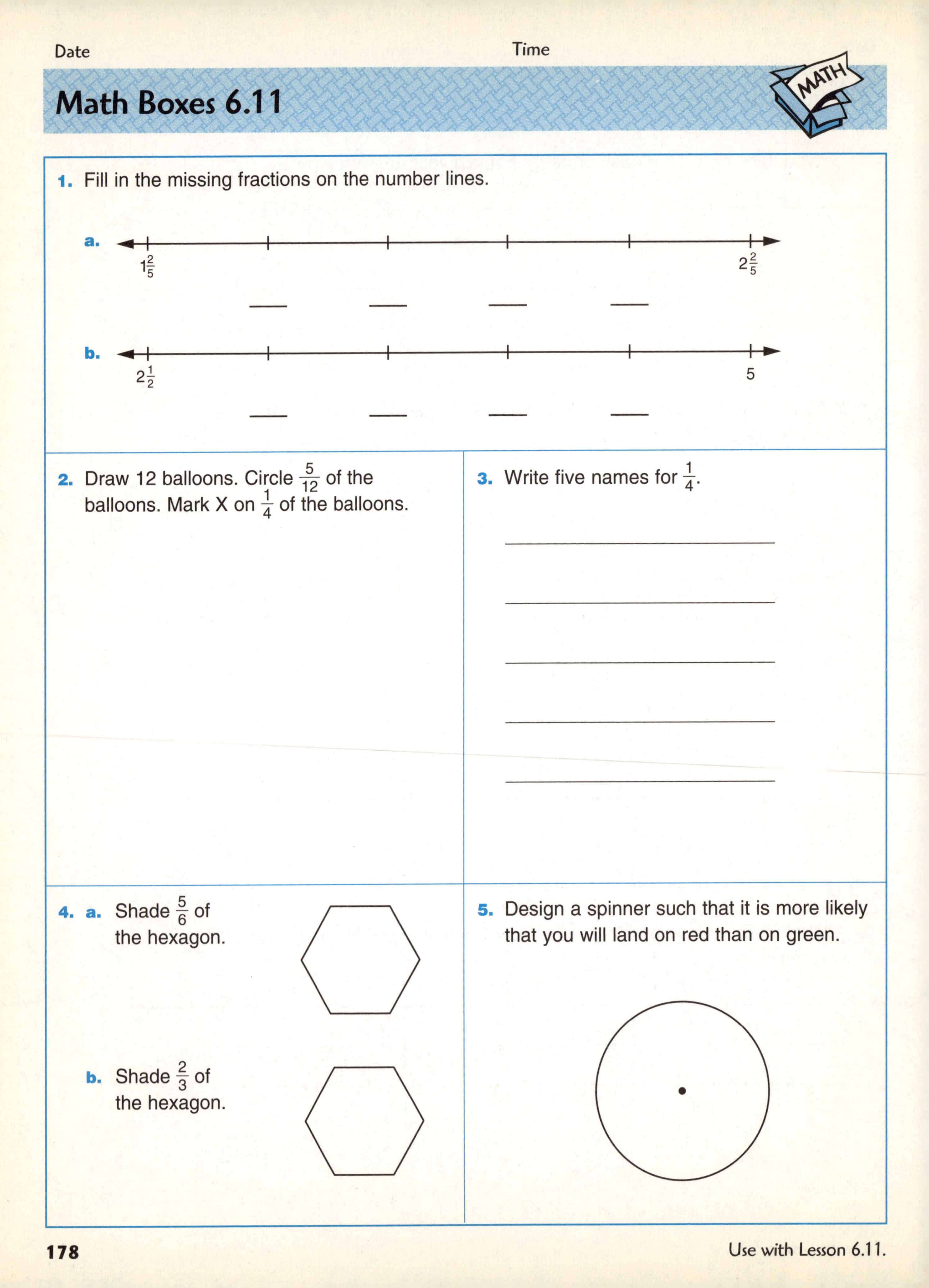

1. Fill in the missing fractions on the number lines.

a. $1\frac{2}{5}$ ___ ___ ___ ___ $2\frac{2}{5}$

b. $2\frac{1}{2}$ ___ ___ ___ ___ 5

2. Draw 12 balloons. Circle $\frac{5}{12}$ of the balloons. Mark X on $\frac{1}{4}$ of the balloons.

3. Write five names for $\frac{1}{4}$.

4. **a.** Shade $\frac{5}{6}$ of the hexagon.

b. Shade $\frac{2}{3}$ of the hexagon.

5. Design a spinner such that it is more likely that you will land on red than on green.

Date Time

My Route Log

Date	Country	Capital	Air distance from last capital	Total distance traveled so far
	1 U.S.A.	Washington, D.C.		
	2 Egypt	Cairo		
	3			
	4			
	5			
	6			
	7			
	8			
	9			
	10			
	11			
	12			
	13			
	14			
	15			
	16			
	17			
	18			
	19			
	20			

Date Time

Route Map

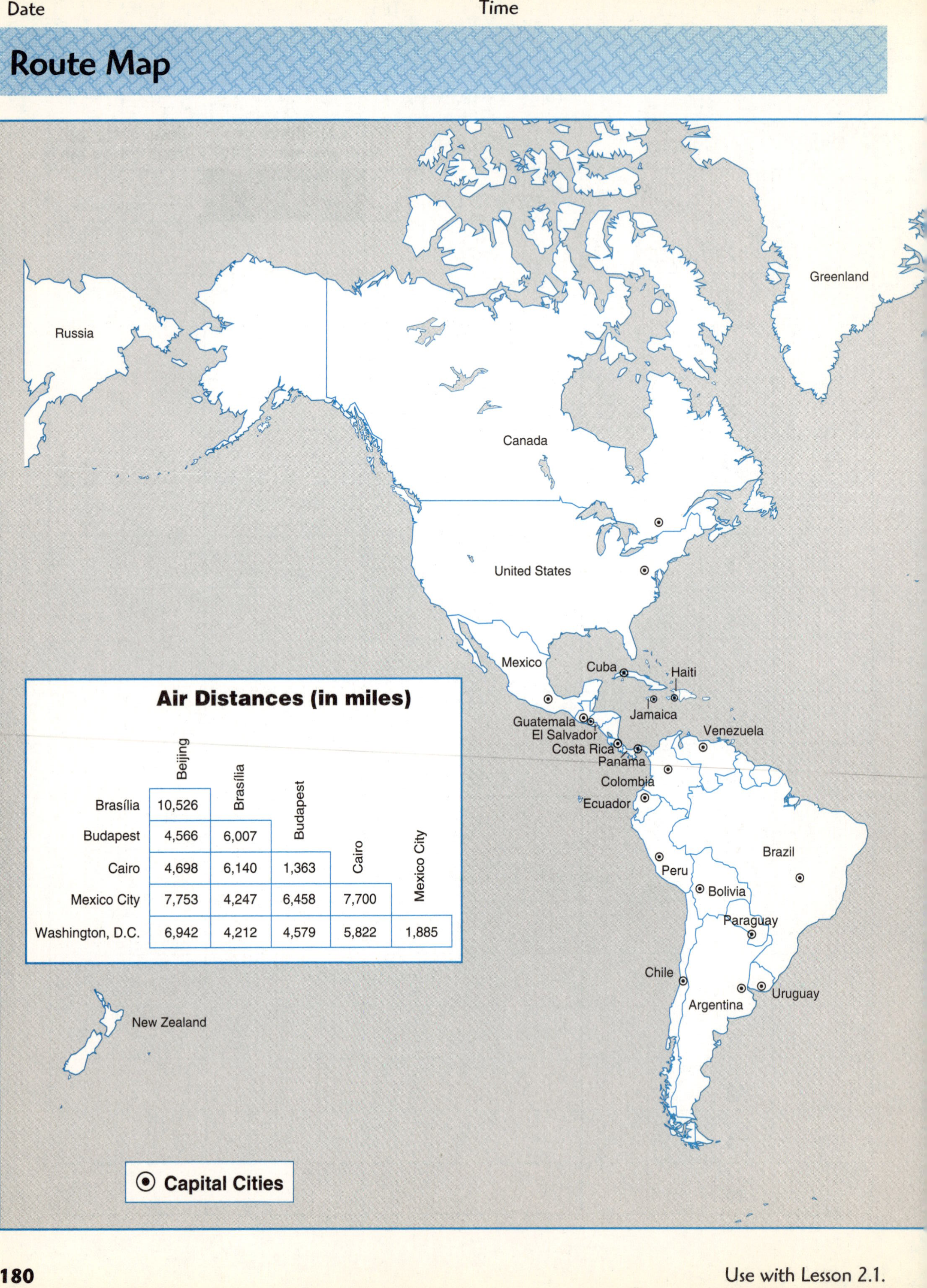

Air Distances (in miles)

	Beijing	Brasília	Budapest	Cairo	Mexico City
Brasília	10,526				
Budapest	4,566	6,007			
Cairo	4,698	6,140	1,363		
Mexico City	7,753	4,247	6,458	7,700	
Washington, D.C.	6,942	4,212	4,579	5,822	1,885

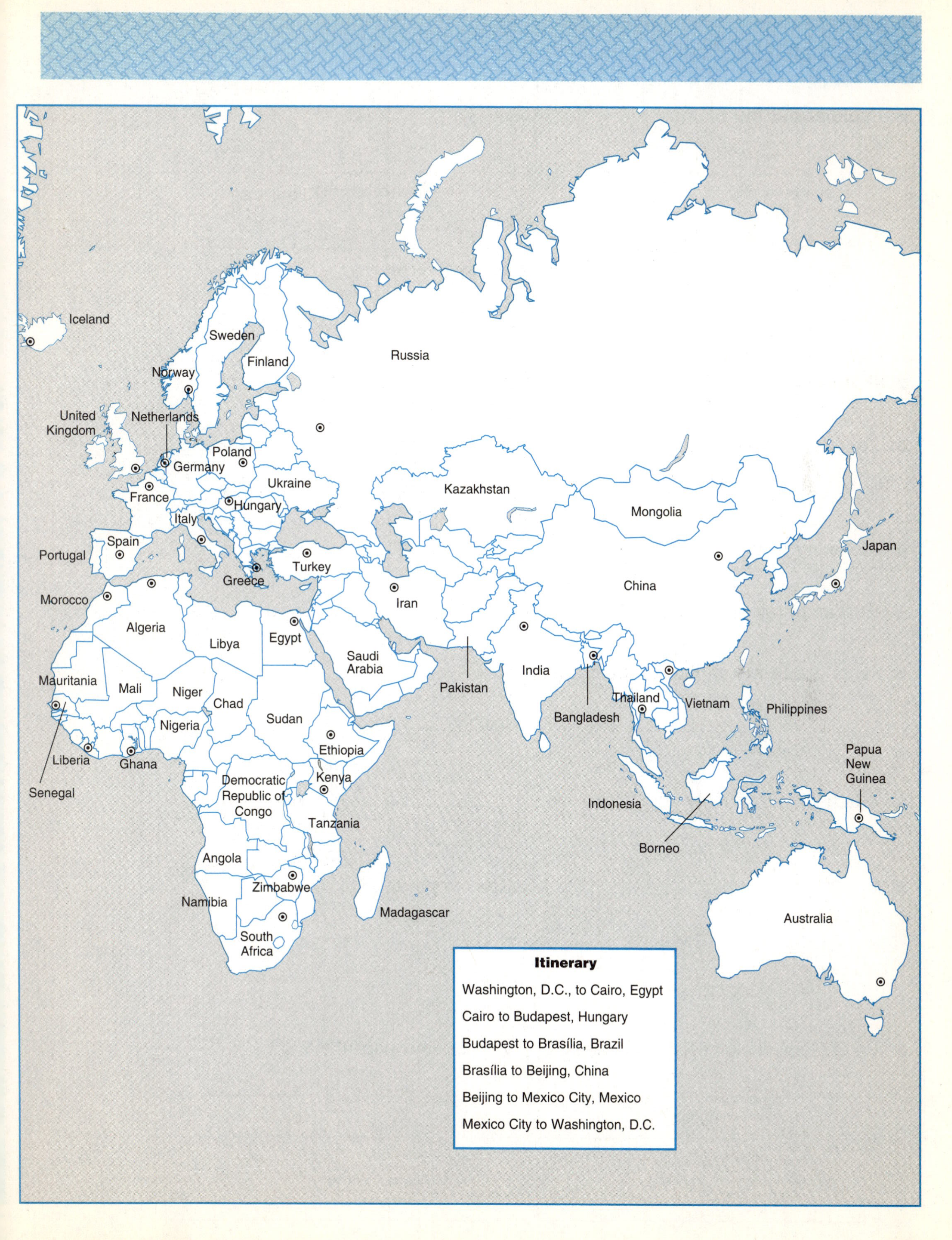
Iceland
Sweden
Finland
Russia
Norway
United Kingdom
Netherlands
Poland
Germany
Ukraine
France
Hungary
Kazakhstan
Mongolia
Italy
Spain
Portugal
Turkey
Greece
Japan
China
Morocco
Iran
Algeria
Libya
Egypt
Saudi Arabia
India
Pakistan
Mauritania
Mali
Niger
Chad
Thailand
Vietnam
Bangladesh
Philippines
Sudan
Nigeria
Ethiopia
Liberia
Ghana
Senegal
Democratic Republic of Congo
Kenya
Papua New Guinea
Indonesia
Tanzania
Borneo
Angola
Zimbabwe
Namibia
Madagascar
Australia
South Africa
Itinerary
Washington, D.C., to Cairo, Egypt
Cairo to Budapest, Hungary
Budapest to Brasília, Brazil
Brasília to Beijing, China
Beijing to Mexico City, Mexico
Mexico City to Washington, D.C.

Date Time

My Country Notes

A. Facts about the country

__________ is located in __________.
name of country / name of continent

1. It is bordered by __________
countries, bodies of water
__________.

2. Population: __________ Area: __________ square miles

3. Languages spoken: __________
____

4. Monetary unit: __________

5. Exchange rate (optional): 1 __________ = __________

B. Facts about the capital of the country

__________ Population: __________
name of capital

1. When it is noon in my hometown, it is __________ in __________.
time (A.M. or P.M.?) / name of capital

2. In __________, the average temperature in
month
__________ is about __________ °F.
name of capital

3. What kinds of clothes should I pack for my visit to this capital? Why?

____
____
____
____

Date Time

My Country Notes (cont.)

4. Turn to the Route Map found on journal pages 180 and 181.
   Draw a line from the last city you visited to the capital of this country.

5. If your class is using the Route Log, take journal page 179 or *Math Masters,* page 38 and record the information.

6. Can you find any facts on pages 246–249 in your *Student Reference Book* that apply to this country? For example, is one of the 10 tallest mountains in the world located in this country? List all the facts you can find.

____

____

____

____

____

C. My impressions about the country

Do you know anyone who has visited or lived in this country? If so, ask that person for an interview. Read about the country's customs and about interesting places to visit there. Use encyclopedias, travel books, the travel section of a newspaper, or library books. Try to get brochures from a travel agent. Then describe below some interesting things you have learned about this country.

____

____

____

____

____

____

____

____

____

Date ______ Time ______

My Country Notes

A. Facts about the country

______ is located in ______.
name of country — name of continent

1. It is bordered by ______
countries, bodies of water
______.

2. Population: ______ Area: ______ square miles

3. Languages spoken: ______
______

4. Monetary unit: ______

5. Exchange rate (optional): 1 ______ = ______

B. Facts about the capital of the country

______ Population: ______
name of capital

1. When it is noon in my hometown, it is ______ in ______.
time (A.M. or P.M.?) — name of capital

2. In ______, the average temperature in
month
______ is about ______ °F.
name of capital

3. What kinds of clothes should I pack for my visit to this capital? Why?
______
______
______
______

Date Time

My Country Notes (cont.)

4. Turn to the Route Map found on journal pages 180 and 181.
Draw a line from the last city you visited to the capital of this country.

5. If your class is using the Route Log, take journal page 179 or *Math Masters,* page 38 and record the information.

6. Can you find any facts on pages 246–249 in your *Student Reference Book* that apply to this country? For example, is one of the 10 tallest mountains in the world located in this country? List all the facts you can find.

____
____
____
____
____

C. My impressions about the country

Do you know anyone who has visited or lived in this country? If so, ask that person for an interview. Read about the country's customs and about interesting places to visit there. Use encyclopedias, travel books, the travel section of a newspaper, or library books. Try to get brochures from a travel agent. Then describe below some interesting things you have learned about this country.

____
____
____
____
____
____
____
____
____
____

Date Time

My Country Notes

A. Facts about the country

________________ is located in ________________________.
name of country name of continent

1. It is bordered by ________________________________
countries, bodies of water
____.

2. Population: ____________ Area: ____________ square miles

3. Languages spoken: ______________________________
____

4. Monetary unit: ______________________________

5. Exchange rate (optional): 1 ____________ = ____________

B. Facts about the capital of the country

________________ Population: ________________
name of capital

1. When it is noon in my hometown, it is __________ in ____________.
time (A.M. or P.M.?) name of capital

2. In ________________, the average temperature in
month
________________ is about __________ °F.
name of capital

3. What kinds of clothes should I pack for my visit to this capital? Why?
____
____
____
____

Date Time

My Country Notes (cont.)

4. Turn to the Route Map found on journal pages 180 and 181. Draw a line from the last city you visited to the capital of this country.

5. If your class is using the Route Log, take journal page 179 or *Math Masters,* page 38 and record the information.

6. Can you find any facts on pages 246–249 in your *Student Reference Book* that apply to this country? For example, is one of the 10 tallest mountains in the world located in this country? List all the facts you can find.

C. My impressions about the country

Do you know anyone who has visited or lived in this country? If so, ask that person for an interview. Read about the country's customs and about interesting places to visit there. Use encyclopedias, travel books, the travel section of a newspaper, or library books. Try to get brochures from a travel agent. Then describe below some interesting things you have learned about this country.

Date Time

*, / Fact Triangles 1

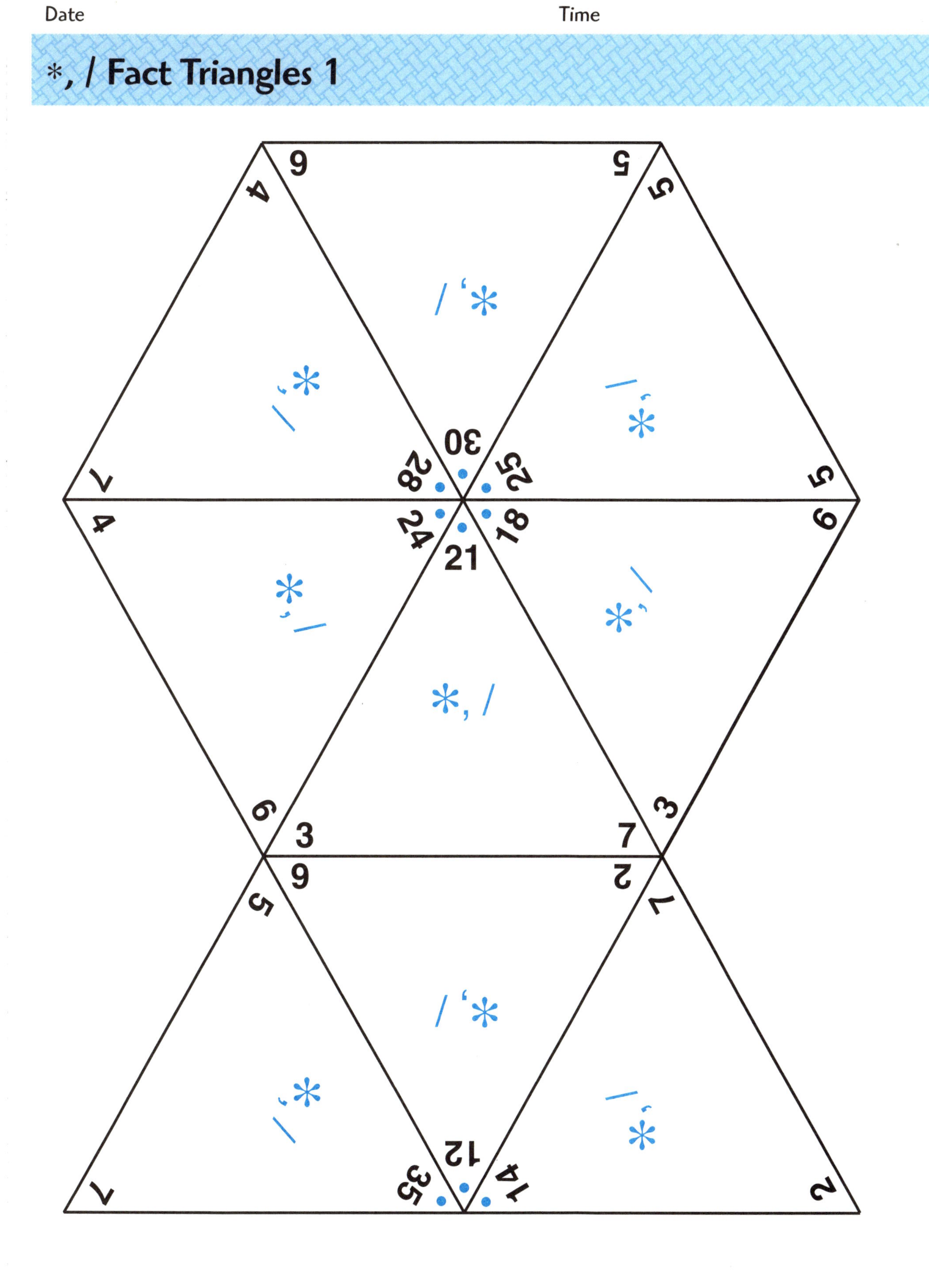

Date Time

*, / Fact Triangles 2

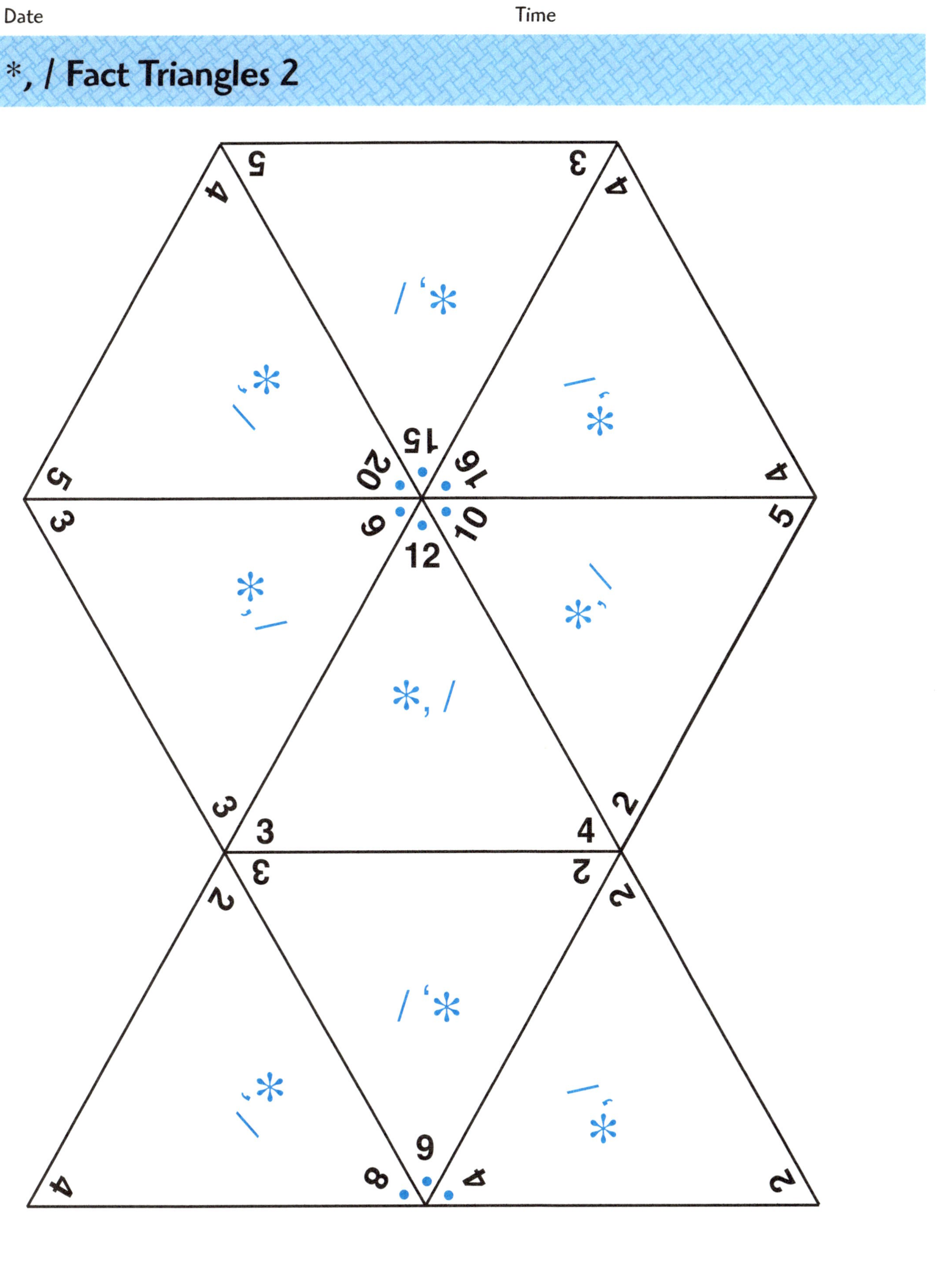

Date Time

*, / Fact Triangles 3

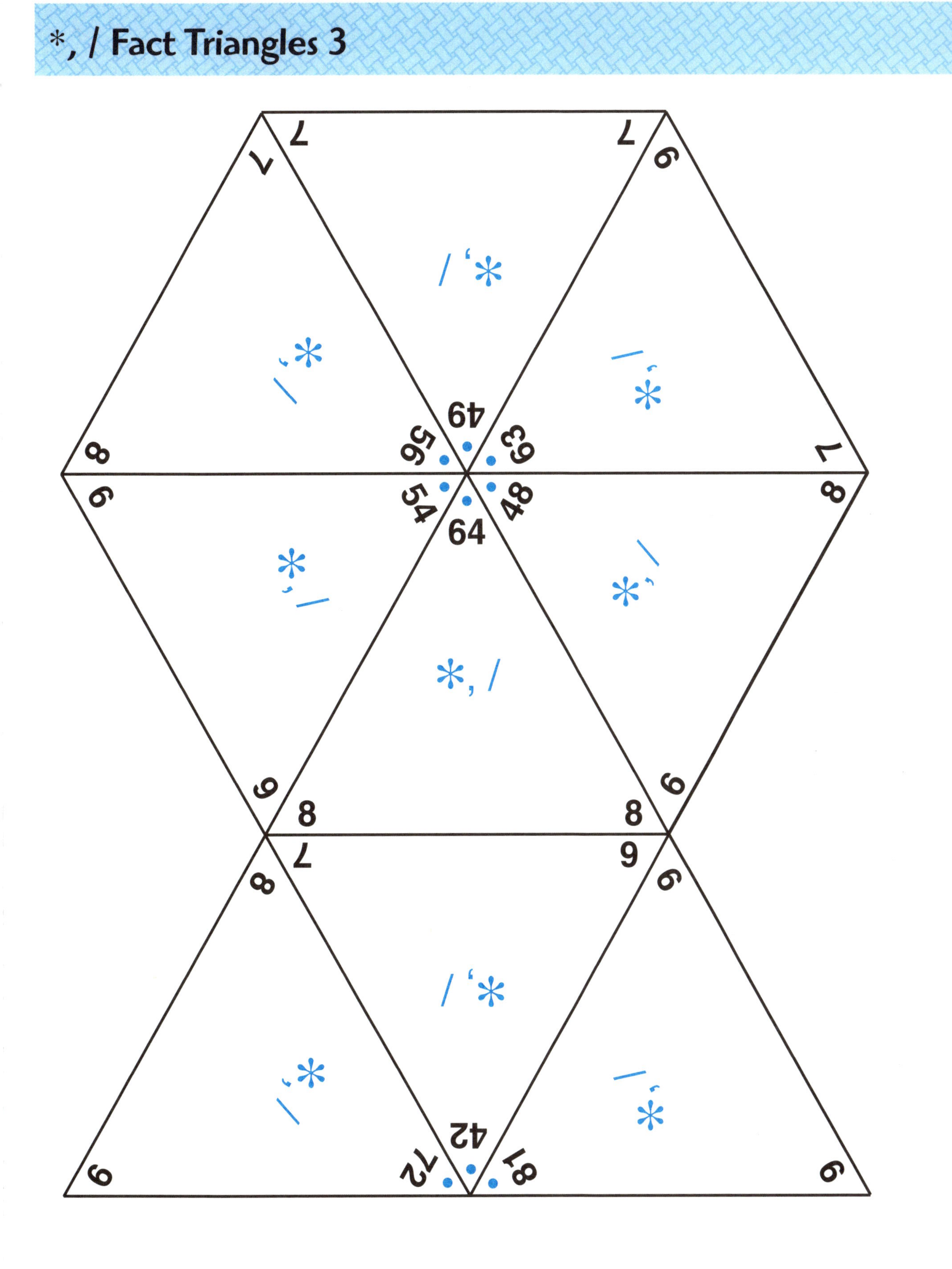

Date Time

*, / Fact Triangles 4

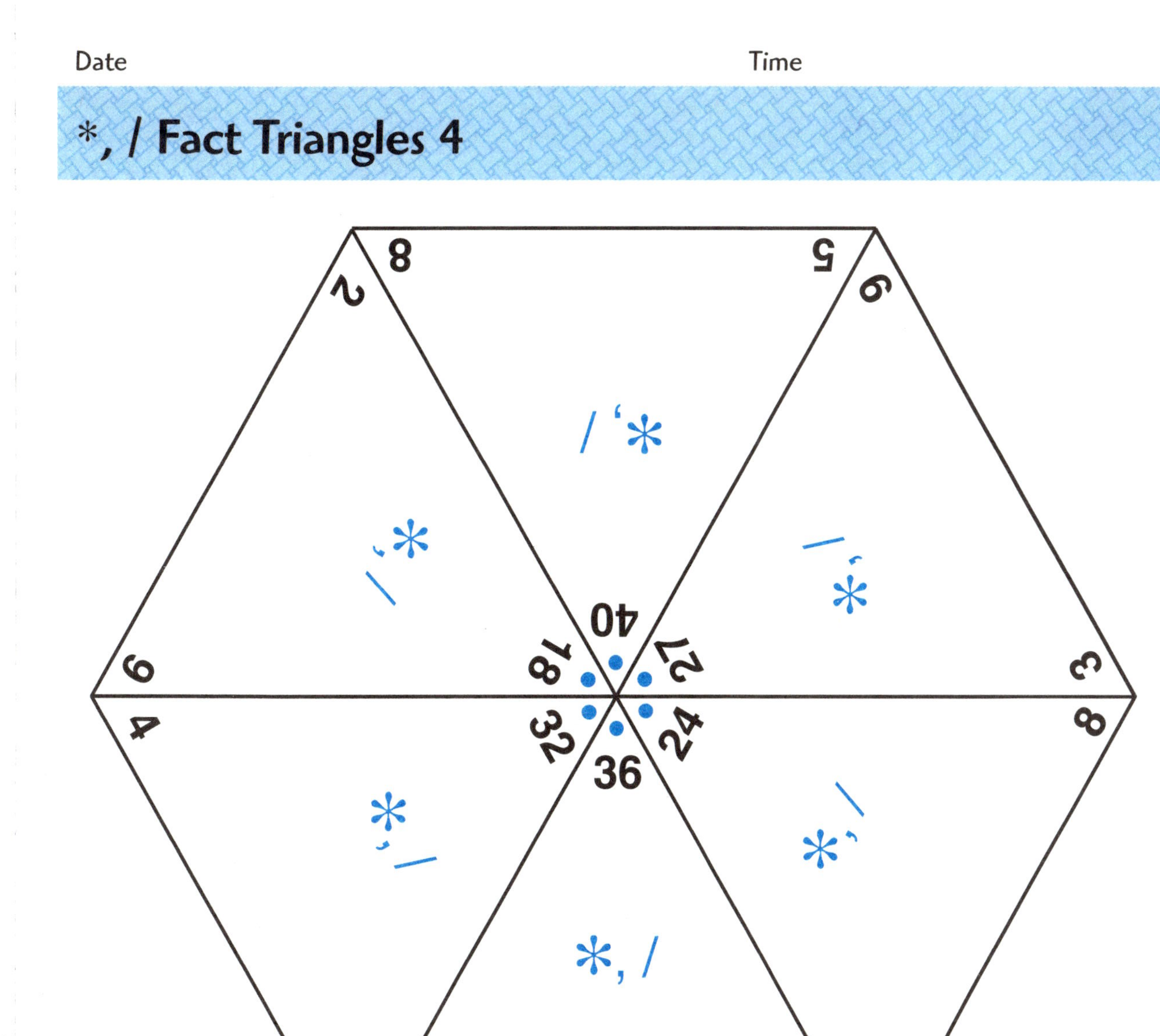